Main Break Prediction, Prevention, and Control

Awwa Research Foundation

About the Awwa Research Foundation

The Awwa Research Foundation (AwwaRF) is a member-supported, international, nonprofit organization that sponsors research to enable water utilities, public health agencies, and other professionals to provide safe and affordable drinking water to consumers.

The Foundation's mission is to advance the science of water to improve the quality of life. To achieve this mission, the Foundation sponsors studies on all aspects of drinking water, including supply and resources, treatment, monitoring and analysis, distribution, management, and health effects. Funding for research is provided primarily by subscription payments from approximately 1,000 utilities, consulting firms, and manufacturers in North America and abroad. Additional funding comes from collaborative partnerships with other national and international organizations, allowing for resources to be leveraged, expertise to be shared, and broad-based knowledge to be developed and disseminated. Government funding serves as a third source of research dollars.

From its headquarters in Denver, Colorado, the Foundation's staff directs and supports the efforts of more than 800 volunteers who serve on the board of trustees and various committees. These volunteers represent many facets of the water industry, and contribute their expertise to select and monitor research studies that benefit the entire drinking water community.

The results of research are disseminated through a number of channels, including reports, the Web site, conferences, and periodicals.

For subscribers, the Foundation serves as a cooperative program in which water suppliers unite to pool their resources. By applying Foundation research findings, these water suppliers can save substantial costs and stay on the leading edge of drinking water science and technology. Since its inception, AwwaRF has supplied the water community with more than $300 million in applied research.

More information about the Foundation and how to become a subscriber is available on the Web at **www.awwarf.org.**

Main Break Prediction, Prevention, and Control

Prepared by:
Neil S. Grigg
Colorado State University
Department of Civil and Environmental Engineering, Fort Collins CO 80523

Sponsored by:
Awwa Research Foundation
6666 West Quincy Avenue, Denver, CO 80235-3098

Published by:

Awwa Research Foundation

American Water Works Association

IWA Publishing

DISCLAIMER

This study was funded by the Awwa Research Foundation (AwwaRF). AwwaRF assumes no responsibility for the content of the research study reported in this publication or for the opinions or statements of fact expressed in the report. The mention of trade names for commercial products does not represent or imply the approval or endorsement of AwwaRF. This report is presented solely for informational purposes.

CONTENTS

LIST OF TABLES ix

FOREWORD xi

PREFACE xiii

ACKNOWLEDGMENTS xv

EXECUTIVE SUMMARY xvii

CHAPTER 1: INTRODUCTION, BACKGROUND, AND SCOPE OF REPORT 1
- Why Utilities Must Anticipate Main Breaks 1
- Purpose of the Report 2
- Original Project Background 3

CHAPTER 2: PRIOR RESEARCH ON PREDICTION OF MAIN BREAKS 5
- Purpose of Chapter 5
- Prioritization of Pipe Renewal 5
- Formulating the Main Break Problem 5
- Research on Pipe Failure Causes and Prediction 7
- Aggregated Models for Systemwide Planning 14
- Ongoing AwwaRF Research 14

CHAPTER 3: MAINS BREAK PREVENTION SURVEY 17
- About the Original Project Survey 17
- Findings 17
  - Inventory and break rates 17
  - Expected value of research to utilities 17
  - Priority-setting 18
  - Current practices in prioritization 18
  - Data availability 18
  - Consequences 19
- Comparing this Survey to Other Surveys 19
- Conclusions About Project Survey 21

CHAPTER 4: LESSONS FROM MODEL DEVELOPMENT AND TESTS 23
- Purpose of Chapter 23
- Assumptions for Model Development 23
- Data Collection and Preparation for the Model 23
  - Characteristics of mains 23
  - Experience in collecting data 24
  - How to handle data 24
  - Data from utilities 25

Approximations of main attributes 26
Cost data 27
Model Development 28
Concepts 28
Dependent variables 29
Data limits 29
Questions to answer by modeling 29
Past models and promising methods 29
Data limitations 30
Modeling Scope Focus 30
Results and elimination of less promising approaches 30
Break types and time intervals 32
Functional forms of the models 32
Development of the models 33
Stage 1 Results (Breaks or no breaks) 33
Multiple Breaks Models 34
Feasibility of Segment Level predictions 35
Significance of different periods and types of break history 35
Models by material type 35
Preliminary conclusions of modeling methods 35
Model Applied to Utility Data 35
Some Conclusions about Modeling 36
Main Replacements Justification 36
Definitions of mains segments 37
Case Studies 38
Study descriptions 38
Understanding and usability 39
Breaks data preparation and loading 39
Prioritization report 40
Case studies summary 40
Summary of Chapter 40

CHAPTER 5: SUMMARY AND CONCLUSIONS 43
Why Utilities Need to Anticipate Main Breaks 43
Project Background and Status of Main Break Research 43
Prioritization of Water Main Renewal 44
Data Management 44
Mathematics of Main Break Histories 44
Risk Factors for Main Breaks 44
Human Factors and Utility Acceptance of Models 45
Overall Conclusions and Future Research 45

CHAPTER 6: RECOMMENDATIONS TO UTILITIES AND AwwaRF 47
Utility Risk Analysis in Replacement Planning 47
How Utilities Can Perform Risk Analysis for Main Breaks 47
Water Supply Industry Research 48

APPENDIX: SURVEY RESULTS AND QUESTIONS 51

REFERENCES 59

ABBREVIATIONS 63

TABLES

3.1 Utility practices in replacement planning....19

3.2 Inventory of distribution systems....19

3.3 Causes of main breaks....20

3.4 Criteria for pipe replacement....20

3.5 Comparison of distribution system data....21

4.1 Comparison of collected inventory data and break rates....25

4.2 Cost data on main breaks....28

4.3 Characteristics of case study utilities....28

FOREWORD

The Awwa Research Foundation is a nonprofit corporation that is dedicated to the implementation of a research effort to help utilities respond to regulatory requirements and traditional high-priority concerns of the industry. The research agenda is developed through a process of consultation with subscribers and drinking water professionals. Under the umbrella of a Strategic Research Plan, the Research Advisory Council prioritizes the suggested projects based upon current and future needs, applicability, and past work; the recommendations are forwarded to the Board of Trustees for final selection. The foundation also sponsors research projects through the unsolicited proposal process; the Collaborative Research, Research Applications, and Tailored Collaboration programs; and various joint research efforts with organizations such as the U.S. Environmental Protection Agency, the U.S. Bureau of Reclamation, and the Association of California Water Agencies.

This publication is a result of one of these sponsored studies, and it is hoped that its findings will be applied in communities throughout the world. The following report serves not only as a means of communicating the results of the water industry's centralized research program but also as a tool to enlist the further support of the nonmember utilities and individuals.

Projects are managed closely from their inception to the final report by the foundation's staff and large cadre of volunteers who willingly contribute their time and expertise. The foundation serves a planning and management function and awards contracts to other institutions such as water utilities, universities, and engineering firms. The funding for this research effort comes primarily from the Subscription Program, through which water utilities subscribe to the research program and make an annual payment proportionate to the volume of water they deliver and consultants and manufacturers subscribe based on their annual billings. The program offers a cost-effective and fair method for funding research in the public interest.

A broad spectrum of water supply issues is addressed by the foundation's research agenda: resources, treatment and operations, distribution and storage, water quality and analysis, toxicology, economics, and management. The ultimate purpose of the coordinated effort is to assist water suppliers to provide the highest quality of water economically and reliably. The true benefits are realized when the results are implemented at the utility level. The foundation's trustees are pleased to offer this publication as a contribution toward that end.

Walter J. Bishop
Chair, Board of Trustees
Awwa Research Foundation

Robert C. Renner, P.E.
Executive Director
Awwa Research Foundation

PREFACE

This report presents a summary of work completed under the AwwaRF project titled, "Main Break Prediction, Prevention, and Control," which was conducted by Fred Reid of URS Corporation starting in 1997. The original project objective focused on cause and consequence factors for risk from main breaks and leaks, a methodology to assign risk factors to individual pipes, and specific information about risk variables. The project was not finished on its original schedule, but useful research was accomplished and this is now the final report for the project. Part of the project effort was devoted to preparing software for main renewal planning, but the software was not finished and could not be included in this report. This report includes a review of literature about main break prediction, results of a utility survey, a summary of work accomplished on the project's modeling objectives, and updated information about predicting main breaks. Utilities can use the information in this report to learn the status of main break modeling and data management and to gain confidence in their own methods to assess risk of main breaks and plan for renewal. AwwaRF can use the information to help plan its future research agenda.

ACKNOWLEDGMENTS

The project's participating utilities and their representatives deserve the most credit for informing the investigators about their practices and in guiding the research. The authors of this report are indebted to the following water utilities and individuals for their cooperation and participation in the project. Fred Reid of URS Corporation was the original project manager. San Jose Water Company was a financial contributor to the project. Participating utilities were: City of Houston – Department of Public Works, St Louis Water Division, NYCDEP, Newport News Department of Water, and EBMUD. San Jose Water Company is the financial contributing utility. PAC members were: Graeme Burns, NLTRC-Ondeo; Dennis Mahoney, Aqua America, Inc.; Ken Mueller, Missouri-American Water Company; and Stuart Tilley, NLTRC-Ondeo.

EXECUTIVE SUMMARY

WHY UTILITIES SHOULD ASSESS RISK OF MAIN BREAKS

As reported by industry surveys, water utilities in the United States suffer some 250,000 to 300,000 main breaks per year. These will cause total annual damages and indirect consequences of about $3 billion, according to the average losses reported by this project's survey. To avoid or mitigate these losses, utilities should assess the condition and risk and make capital decisions about repair, rehabilitation, and renewal of individual pipes.

When utilities prepare business cases for pipe renewal in their capital budgets, they assess the risk of main failures explicitly or implicitly so they can estimate the return on investment of renewal projects. Unfortunately, it is difficult for them to calculate return on investment with confidence and to decide which pipes to renew because deterioration and exposure of pipes take place out of view and utilities cannot accurately assess condition or predict probability of failure of mains. To avoid renewing a pipe too early, many utilities wait for problems to develop before investing in rehabilitation or replacement. Therefore, renewal of water mains in the United States proceeds at a slow rate.

This report presents a summary of work completed under AwwaRF Project 461, "Main Break Prediction, Prevention, and Control," which was initiated in 1997. The original project objective focused on cause and consequence factors for risk from main breaks and leaks, a methodology to assign risk factors to individual pipes, and specific information about risk variables. The project was not finished on its original schedule, but useful research was accomplished and this is now the final report for the project. Part of the project effort was devoted to preparing software for main renewal planning, but the software was not finished and is therefore not included in this report. This report includes a review of literature about main break prediction, results of a utility survey, a summary of work accomplished on the project's modeling objectives, and updated information about predicting main breaks. Utilities can use the information in this report to learn the status of main break modeling and data management and to gain confidence in their own methods to assess risk of main breaks and plan for renewal. AwwaRF can use the information to help plan its future research agenda.

RESEARCH REVIEW ON MAIN BREAK PREDICTION

Project 461 included a literature review to assess factors contributing to breaks, methods of accounting for break consequences, data requirements and sources (mains and consequences), existing applications of models, and technology transfer from gas distribution and transmission pipelines. Since the project ended, additional studies and papers have considered similar issues and are reviewed in this report.

Data and reports on reasons for pipe failure began to appear after the 1940s. After the 1960s, changes in pipe materials and methods added to the complexity of the prediction problem. In the last three decades, only a few research teams have reported on main break causes and prediction. Most reports present analytical ideas, but lack data bases of actual break records. A number of different mathematical functions have been fit to main break data, but the datasets lack the integrity needed to yield consistent results.

Research on main break prediction and planning began during the 1980s when several studies compared analytical predictions to main break data in a few utilities. The studies showed that breaks in the early stages of main life follow different predictor functions than those in later stages do, and the graph that shows this is named the "bathtub function" because you have some risk early in a main's life, the risk declines, and then it rises again. This is intuitive because some pipes may fail early due to material or construction faults and then, once they establish a record of performance, they last a long time until they reach old age. This is similar to a car, for example, which might be a "lemon" and fail early, but if it does not, it will last until old age, when it begins to fail over and over again, in spite of repairs.

The literature also showed that databases to verify prediction models are difficult to organize and use. This is logical because no standard to organize and manage the variables exists and because data of main break variables are complex and hard to measure, and the institutional structures needed to ensure integrity in data management are not in place.

To summarize the literature, renewing buried pipes is one of the many competing items in water utilities' capital budgets. To justify renewal, utilities require an estimate of probability of failure of individual pipe segments; otherwise, they have difficulty convincing their boards to invest funds in renewal. Utilities and AwwaRF recognize this, and have made efforts to develop models to predict main breaks. These efforts have increased the industry's understanding of main break phenomena and identified more research needs, but they have not yet produced a reliable analytical method to predict breaks in individual water mains.

SURVEYS OF UTILITY MAIN BREAK EXPERIENCES

Project 461 included a utility survey of main breaks experience and control. The survey was conducted in 1998 so the findings are now some 8 to 9 years old. While the survey findings are dated, they are consistent with other surveys and statistical summaries of distribution systems.

Representative findings from the survey are:

- Average annual break rate (for all materials) of .25 breaks per mile per year
- Cast iron break rates are significantly higher than other materials
- Utilities rate breaks are most important criteria in prioritizing replacements
- Utilities vary in the methods used to select mains for replacement
- Utilities see assessing mains life as a clear issue but assessing costs is a lesser problem
- Utilities prefer that research focus on segments rather than categories of mains
- Only a few utilities have tried breaks prediction methods
- Availability of consequence data by utilities is very limited

PROJECT FINDINGS ABOUT MODELING MAIN BREAKS

The project included significant effort to develop and test main break models. Efforts included data collection, data cleaning and quality control, verification, and application to utility case studies. The efforts clarified that the link between main break models and risk factors is in the predictor variables that show risk of breaks. Examples of predictor variables are pipe age, diameter, material, soil corrosivity, and pressure, among others. General knowledge within the water industry shows that age of main is only a minor factor in the later stages of failure, the stage when most replacement decisions are made. Again, this is intuitive and is like an old house, car, or machine. The house, car, or machine might be old, but many factors other than age determine if it will fail in the next few years.

The project was able to model some types of breaks in single utilities with a fair degree of confidence, but many questions remained about usefulness of results within the utilities. Usually, different coefficients were required to model different scenarios and pipe variables, even within a single utility. The model exercises in utilities were complex, more like research projects than like routine applications of ready-to-use software. There were some promising results, but many questions remained about data, model estimation (meaning to fit the models to the data), and use of the software by utility personnel.

The project was not able to develop coefficients for predictor variables that can be transferred from one utility to another because there is too much variation in the data. If risk factors were the same across different utilities, then a generalized model could be developed. However, they are not the same, and pooled data from different utilities cannot be used to predict main breaks. In the future, a nationwide data base may show interesting information about risk factors that can be used to inform design and construction.

While pooled data from different utilities cannot be used to create coefficients for models, a utility can use its own pooled data to replace missing data for individual main segments. The project developed a method to accomplish this and explained the restrictions on use of the method to avoid bias in modeling. For example, if you were missing observations of certain mains characteristics to match break records, you could use corresponding data from the utility's pooled data to fill in the missing information. The hybrid method used for this is titled the "cross-classified attribute record." The project also produced guidance for quality assurance of main break data.

The project showed that main break modeling is complex, and requires trained modelers to develop and run the models. Each utility must use its own data to calibrate its model, and such efforts would require statistical expertise and could be costly, even if data are available. Considerable effort is required to perform quality assurance on the data. Even when models are calibrated, confidence levels in the statistical estimates are often lower than needed to give credibility to model results for use in justifying pipe renewal investments. Utilities are wary of using "off-the-shelf" software until it is thoroughly proven, and the large investments required for pipe renewal require solid justification in budget processes, not predictions based on models without thorough testing and validation. Better ways must be found to explain main break data and models to managers and employees within utilities who would implement them.

In conclusion, neither Project 461 nor other studies have led to a reliable model for main failure prediction, but they did break ground in main break research and laid the basis for future models. While models for breaks in individual pipes have not worked well yet, models for aggregated pipe systems offer useful methods to compile information about network condition and to establish longer term capital improvement programs for systems.

PLANNING FOR WATER MAIN RENEWAL

The most important reason for calculating risk of main breaks is to plan and prioritize renewal projects so as to mitigate risk. While utilities recognize this, they are unable to compute risk with confidence and their renewal rates for aging infrastructure are low because they take a risk-averse approach to decision making. They lack data to prove the need for more aggressive approaches, and they desire to keep rate increases low.

As a matter of fact, this conservative strategy for mains renewal has not been proved to be incorrect and, in fact, some utilities believe that a very low or reactive replacement rate, as in Australia, may be a correct investment strategy. Therefore, improving the process of risk assessment for main breaks is necessary to identify the correct policy for mains renewal, as well as to prevent and mitigate losses from main breaks.

In any event, to gain confidence in their decisions about renewal rates, utilities should perform risk analysis to anticipate main breaks, prevent direct losses, and alleviate customer and public losses. To know how to do this, they require convincing business cases for pipe renewal, and these require that utilities anticipate failures of individual water mains. To predict failure requires an effective model, even if the model is a judgment–based assessment by experienced personnel.

Although they cannot use models to predict pipe failure with confidence, some utilities use point systems based on data and/or judgment to assign priorities. This approach has merit, but is less satisfactory than knowing probability of failure, which will improve the preparation of business cases. Regardless, both point systems and failure probability models rely on knowledge of pipe risk factors, such as age, material, and others.

While point systems and decision matrices are not perfect, they do represent the state of the art of decision analysis for utility renewal decisions. Break rates can be used within point systems to provide a proxy measure of risk. A utility can use the research in this report to inform its engineers and researchers about the experience of other utilities in using break rates to schedule pipe renewal program. This utility would then need to develop and test methods such as decision matrices for application to its own system. In addition to break rates and risk factors, utilities should consider consequences of failure to drive their decisions. Utilities can also consider opportunities, such as to replace problematic pipe when streets are re-paved.

To pinpoint needs for main break analysis, utilities can study the decisions to allocate funds in capital improvement planning and to schedule renewal in individual pipes. Capital allocation can be done by tracking main breaks across the network and computing return on investment of pipe renewal. The individual mains problem is like a medical situation where failure of a component can involve a potential disaster, and it pays to invest to determine specifically which mains are most at risk. This analysis can be done by modeling or by sorting mains by risk factors using simpler methods. The key is a consistent and coherent data base.

For the future, individual utilities can develop learning methods to identify risk factors, and for each new break they can add new lessons to their data bases. They also can train personnel for analysis of distribution systems. They can initiate data collection and analysis programs, including main breaks research programs based on data.

WATER SUPPLY INDUSTRY RESEARCH

Main break research has proceeded for over twenty years, and emphasis has been on both modeling and data management. However, without adequate data protocols and standards, further model efforts will be of limited use. Until water main break data bases become more reliable and standardized, utilities will not have confidence in main break prediction models. Therefore, the focus in the next stages of research should be on datasets to predict main failures. Research should focus on uses of the data, a protocol, standard methods and quality control for main break data. After data management methods are developed and standardized, the next issue will be to create incentives for utilities to collect and manage the data.

CONCLUSIONS

Nationally, water utilities suffer some 250,000 to 300,000 main breaks that cause total losses of about $3 billion per year. To mitigate these, utilities require effective strategies to assess condition, compute risk, and implement pipe renewal projects. Utilities are renewing their mains very slowly, once in 200 years or less. Planning for mitigation requires valid estimates of probability of failure, and providing these estimates has been the goal of a series of research projects, including Project 461. This report provided information about this research, including a review of literature, results of a survey, a summary of work accomplished on modeling, and additional information about main breaks.

The slow rate of renewal shows that most utilities are not convincing their boards to invest funds in asset renewal. Research efforts have increased the industry's understanding of main break phenomena and identified more research needs, but they have not yet produced a reliable method to assess risk to individual pipes. While models for individual pipes have not worked well yet, models for aggregated pipe systems do offer useful methods to compile information and to establish longer term capital improvement programs for systems.

Meanwhile, once they have obtained their renewal budgets, utilities can use point systems and decision matrices to consider risk factors, consequences, and opportunities to select pipes for renewal. The problem remains to justify the renewal budgets.

For the future, utilities can develop learning methods to identify risk factors, organize their data bases, and train their personnel. Industry research should focus on a protocol, standard methods and quality control for main break data. After these are available, the next issues will be incentives and guidance for utilities to adopt them.

CHAPTER 1
INTRODUCTION, BACKGROUND, AND SCOPE OF REPORT

WHY UTILITIES MUST ANTICIPATE MAIN BREAKS

Buried pipes are the most costly capital assets of water utilities, and their condition is a major issue to utility managers (Runge and Mann 2005). In its 2006 survey, AWWA found that utilities are more concerned with infrastructure in 2006 than in 2005, with the adjective "failing" beginning to replace "aging" to describe infrastructure condition, thus indicating increased concern among utilities (Parmelee 2006).

Main breaks are an indicator of infrastructure condition, and utilities in the United States suffer about 250,000 to 300,000 of them every year. Utilities are challenged to control these breaks while managing large inventories of buried assets. Although cost models for main breaks have been prepared, it is not yet possible to make general estimates[1]. However, if each main break causes $5,000 in direct losses to the water utility and another $5,000 in indirect losses and consequences, the cost to the U.S. is about $3 billion per year nationally, or $10 per year per person. The direct losses to the water utility would add on the order of $1.50 per month to the average residential water bill. Indirect costs and consequences cause loss of service, disruption, and loss of customer confidence. In addition, weak water mains cause water loss, reduced pressure and higher energy costs, and other problems.

Taken in perspective, losses due to main breaks are on the same order of magnitude as annual flood losses, which are estimated at more than $2 billion of property damage in the U.S.[2] Flood loss is considered as the most costly natural disaster in the nation. Therefore, water main breaks are a serious problem and require deliberate efforts to mitigate them.

When utilities prepare business cases for pipe replacement in their capital budgets, they must show when breaks in individual pipes will occur to compute rates of return and risk.

If utilities know which pipes are at risk of failing and what the consequences will be, they can plan maintenance strategies, improve reliability, identify pipes for renewal, and justify their capital budgets.

The problem is similar in concept to that of replacing and renewing an aging fleet of vehicles. You know in general what the replacement rate should be to avoid breakdowns and loss of service, but it is a challenge to know which vehicles to repair or replace at any time. The fleet problem is much simpler because assessing condition is much easier for a vehicle than it is for a buried pipe. Also, one segment of a buried pipe can be as much as a mile of pipe with many different elements, whereas one vehicle is a readily-identified management unit, albeit with many different parts.

While surveys show that utilities are aware that they should undertake pipe renewal programs, it is difficult to decide which pipes to renew because deterioration and exposure of them take place out of view. Therefore, it is difficult to assess their condition and to plan renewal programs (Grigg, 2004). Utilities are conservative and customers are cost conscious, so the risk of renewing a pipe that does not require it tends to make utilities wait for problems to develop before renewing pipes. Utilities are currently seeking to understand better how

---

[1] See Deb, et. al., (2002) and Cromwell et. al. (2002) for cost models and explanations.

[2] Flood losses are difficult to estimate. This estimate of losses is taken from the web page of the Federal Emergency Management Agency (2006), but after Hurricane Katrina and floods of 2006, the loss estimates will rise.

customers view the trade offs between renewal, cost, and risk. A recent AwwaRF probed customer acceptance of water main reliability (Damodaran et. al. 2005).

While predicting probability of pipe failure is important to utilities for improving operations and capital management, utilities are not able to predict probability of individual pipe failures with confidence. Thus, utilities face uncertainty in making investment decisions about risk because of the unknown condition of buried pipes. It is easier for utilities to estimate condition of their total networks or of zones than it is to identify condition of individual pipes. Thus, utilities can assess the rate of pipe failure in the aggregate, but this assessment does little good in the decision of which pipes to renew at a given time.

Although they cannot predict pipe failure with confidence, some utilities are using point systems to assign priorities for replacements. This approach has merit, but is less satisfactory than knowing probability of failure, which could lend more rigor to business cases than arbitrary or subjective point systems. Still, point systems and probability models are similar in that both use predictor variables such as age and condition as indicators of pipes at risk. In any case, renewal and replacement of water mains in the United States are proceeding at a slow rate, compared to their expected life cycles.

PURPOSE OF THE REPORT

AwwaRF has anticipated the need to predict main breaks, and to conduct ongoing research to upgrade knowledge and develop tools to assess water main condition. In response to the main break issue, AwwaRF (1997) initiated a project to study the cause and consequence factors for risk from main breaks and leaks and to create a methodology for water utilities to assign risk. The project was not finished on the original schedule, but valuable research was accomplished. This is the final report for the project, and it includes a review of work accomplished on the original project objectives and updated information about predicting main breaks.

AwwaRF subscribers who might use this report include utilities and consulting firms. Also, AwwaRF cooperates with government agencies in advancing water supply research. Within this family of collaborators, two applications of this research seem likely: direct use by utilities of the techniques and findings, and continued development of tools to predict water main failures. This report responds to both goals, direct application of current tools and providing knowledge for further research and development.

In preparing this present report, the assumption is that utilities can use some of the information from the project work of five to seven years ago but that the work must be interpreted in the light of progress since then. Since the original project did not result in a final model that can be distributed and used, much of this report's value must be found in its reporting of progress toward predicting main breaks.

Direct application of break prediction tools would reduce risk and cost by focusing investment on high-risk, high-consequence areas. Because the intended model was not completed, the contribution this report can make toward that goal is to clarify the state of the art of predicting main breaks and the status of main break prediction tools (see Chapters 5 and 6) and to make recommendations to utilities on how to use main break planning and prediction methods that are available.

An assessment of research needs on main break prediction is also provided to help in explaining the need and planning future research on main breaks.

ORIGINAL PROJECT BACKGROUND

The original project's title was "Main Break Prediction, Prevention, and Control." The background on it is drawn from the original RFP and the draft documents of the uncompleted project.

The original RFP objective was: "Identify cause and consequence factors for determining risk from main breaks and leaks. Develop a methodology for water utilities to determine or assign cumulative risk-and time-variable factors to individual pipes, and to assign potential cost and customer service consequences." As justification of the need for the project, the RFP stated: "...there is no consistent approach for a decision-making process that targets investment or operational practices to reduce risk, frequency, and annual cost of bursts." "The investment targeting approach would potentially reduce the risk and costs due to bursts by correctly focusing investment on high-risk, high consequence areas... A single model would also function as a strong public relations tool, for use in explaining expenditures to customers."

These explanations of the need for the research appear still valid, nearly ten years after the RFP was issued. However, the context in which the research would be applied has changed, as will be explained later in the report.

The RFP sought specific information about risk, and included a list of items expected in the final report:

- Lists of potential risk and consequence factors
- Data on pipe stress from soil and water temperature changes
- Guidance on obtaining local risk and consequence information
- A list of methods to assess risk and consequences
- Numerical approaches to assess risk and consequence factors for individual pipes, with guidance on loss estimation
- A limited number of case studies
- A specification for applying the outputs within GIS or other computer-based systems

Some of these objectives were achieved in the research and some were not. In addition to offering these results, the contractor offered to produce a "fully developed replacement decision model, tested at 3 utilities, with software ready for further implementation" (Woodward-Clyde and San Jose Water Company 1997). As a result, much of the project effort was focused on producing a model and software. However, the model was not completed, so this present report distills lessons learned in the model effort.

To place the project in context, this was its original research approach:

• First period	Literature search. Project survey.
• Second period	Data quality control and preparation (over 160,000 records). Development of statistical method to account for interaction of mains attributes without having segment level data. Initial modeling. Study of breaks consequences.
• Third period	Analysis using breaks prediction model. Development of software specifications. Development of framework for test project methods in case studies.
• Fourth period	Break modeling at fourth utility (three utilities previous).

		Initiation of case studies task. Further work on consequence cost data. Conclusions about modeling results.
•	Case studies	Conduct of case studies
•	Final report	Date of draft final report

Although the original objectives were not met, AwwaRF believes that the logic and models developed by the team, as well as their experiences and observations, remain relevant for water utilities and should be made available in a useable form for subscribers. In addition, the work on this project will strengthen water utility capabilities to utilize knowledge about their systems to underpin their asset management programs.

In summary, this present report synthesizes results from the original project, including its literature review, survey, data collection and analysis, and modeling efforts. These have been updated by adding material from current studies to produce a state of the art report on predicting breaks in individual pipes. It brings published research to the present level of knowledge and identifies further research needs on main break prediction, which will remain as part of AwwaRF's continuing research agenda.

CHAPTER 2
PRIOR RESEARCH ON PREDICTION OF MAIN BREAKS

PURPOSE OF CHAPTER

The original project included a literature review on main break prediction and modeling. The purpose of this chapter is to report on that literature review and on the state of knowledge on main break prediction[3]. In addition to the material from the draft project report (URS, 2002), it includes reviews of additional papers and reports.

The original study evaluated 21 papers in terms of factors contributing to breaks, methods of accounting for break consequences, data requirements and sources (mains and consequences), existing applications of models, and technology transfer from gas distribution and transmission pipelines. Since the project another 10 to 20 studies and papers have considered similar issues.

PRIORITIZATION OF PIPE RENEWAL

The literature shows that renewing buried pipes is one item of many in the capital budgets of water utilities and their host local governments. Justifying renewal requires that its benefits compete with other, more visible projects. It is therefore a challenge to justify renewal of hidden assets unless a strong business case can be made that benefits exceed costs (Grigg, 2004).

Capital budgeting begins with capital planning and programming. To plan renewal of distribution pipes you can analyze a single pipe segment, a zone, or an aggregated system with multiple pipe segments and/or zones. However, while a utility makes plans about its overall system, capital budgets require it to know the condition and likelihood of failure of single pipe segments. These will normally be on the order of 1000 to 2000 feet, usually between major connections or intersections with other pipes. These individual segments are major investments, which can be seen if you consider a simple numeric of 1,000 feet of pipe at $100 per foot, making a capital item of $100,000 or about $500,000 per mile.

The slow rate of renewal of distribution pipes results, in part, from the built in conservatism of utilities and local governments in not spending public funds for projects without strong justification. Therefore to have the needed justification, you need a good idea of probability of failure of individual pipe segments.

FORMULATING THE MAIN BREAK PROBLEM

Predicting failure in individual water mains is a challenging technical problem. Networks of buried water mains operating under pressure include pipes of many different sizes, types, histories, and conditions. These networks include large numbers of hardware elements in the form of valves, fittings, service lines and other appurtenances.

When you combine the many factors involved, the problem of predicting a main break requires knowledge of many variables at the same time. It is a multivariate statistical problem

[3] Some of the material in this chapter is taken directly from the project's draft report (URS, 2002).

with uncertainty because of lack of data on pipe condition, operating conditions, and environmental conditions.

Statistical models for pipe breaks are similar to other models that attempt to use a few explanatory factors to account for the uncertainty of other factors and to create most likely explanations for a failure or other event. A medical analogy would be if Patient A had a heart attack. The patient smokes, is overweight, eats too much fat, and doesn't exercise. Moreover, his father had a heart attack at almost the same age. Which risk factor caused the heart attack? Now that he had one heart attack, what is the probability that he will have another in two years? Does it matter if he underwent some repairs (a stent, a bypass, an angioplasty) and changed the risk factors?

The uncertainly in pipe break prediction creates a dilemma because, if utilities are to use their scarce resources effectively, their capital renewal decisions must be based on actual risk of failure based on information about threats, vulnerabilities, and consequences of failure. At present, utilities do not in general know probabilities of failure of individual pipes or how vulnerable they are. Thus, utilities must integrate the knowledge they do have to make estimates and set priorities.

So, to predict when a break will occur, you would consider the pipe's attributes (age, material, etc.) its environmental variables (soil conditions, vehicle loads, temperature, etc.) and its operating conditions (pressure, flows, surges, water quality). Determining which of these is most important in determining risk of break involves statistical analysis because you cannot know the exact situation for any given main at any time.

In statistical terms, the "event" to be predicted is that a break occurs in water main. What constitutes a break must be defined and, as explained later, some researchers define it as when a utility must repair a broken pipe. The event is converted to a random variable by defining either that an event occurs in a time period, say the next year, or defining the variable in terms of time between breaks or number of breaks in a given period.

Another issue is conditional probability; that is, given that break(s) have occurred, what will be the time to the next one or what is the probability that the next one occurs within say one year?

Whatever the formulation of the statistical problem, the challenge is to use predictor variables to predict probability of a break or time to a break. Predictor variables include pipe attributes, environmental conditions, and loading conditions. The studies to be summarized in this chapter have considered many of these variables.

After reviewing a number of references, this sample list of nine categories of predictor variable was compiled (Grigg, 2005):

- Degradation of material, including internal and external corrosion
- Weak joints and poor installation procedures
- Soil displacements
- Frozen soil around pipe
- Construction accidents
- Water temperature extremes
- External imposed loads
- Manufacturing defects
- Pressure surges

The list is not exhaustive and it can be expressed in different forms. If you consider these variables, none of which is easy to measure, the challenge of making a valid prediction of failure probability is seen as formidable.

RESEARCH ON PIPE FAILURE CAUSES AND PREDICTION

Given the history of water distribution pipes, which appeared in the United States in the late 18th Century and began to increase sharply after 1900, it is natural that failures would attract the attention of engineers. Data and reports on reasons for pipe failure began to appear after the 1940s (Wolfe, 1946; Baracos, Hurst, and Legget, 1955; Dolson, 1955; and Arnold, Clark, Remus, and Niemeyer, 1960). Morris's (1967) paper in Journal, AWWA is cited often, and is one of the earliest comprehensive treatments of cast iron pipe failure causes. After the 1960s, changes in pipe materials and methods added to the complexity of the prediction problem. Today, you have many ages, materials, conditions, and other attributes of pipes to consider, making the failure diagnosis and prediction problem more complex.

In the last three decades, only a few research teams have reported on main break causes and prediction. Shamir and Howard (1979) used an exponential model of the increase of breaks with time to find an optimum year to replace a main. The basis for optimality was the sum of the discounted future costs of maintaining a main was greater than cost of replacing it. Reports such as this advanced analytical ideas, but they lacked extensive data bases of actual break records.

Clark et. al. (1982) attempted a comprehensive statistical analysis of factors expected in water main failures. The study included attributes of mains that had no failures as well as those that had failed. It explained why a failure or break must be defined, and it defined a failure as when a utility had to make a repair. It focused on large diameter pipes, excluding most 6 inch and smaller diameter mains. Utility data was from Cincinnati, New Haven, and Kenosha WI.

The study showed that a single functional relationship between failures and factors causing them did not adequately describe main breaks in early stages of main life or in later stages. It concluded that time to first failure was predicted by a linear model of mains attributes and environment[4]. Number of subsequent failures was fit best by nonlinear models of different attributes and time after first failure. The linear models consider independent effects and nonlinear models (with product of factors) consider interacting effects. Their model efforts used six factors: pressure differential, time since first break, land use over main, type of pipe, and low and high corrosivity factor. These factors had a coefficient of determination of $R^2 = 0.47$ in explaining second stage failures[5]. Most break factors had significantly different coefficients for different utilities and age was only a minor factor in second stage breaks.

Using cost data, Walski and Pelliccia (1982) derived an optimal replacement time for mains. Based on data from Binghamton, NY, they derived a break prediction equation and computed time to replace and the critical break rate to signal that time. While results like these sound promising, it is important to understand that there is a large gap between a research study and availability of a tool that utilities have confidence in. They found that age of pipe was a minor factor compared to location and prior leakage. Cast iron mains were up to 100 years old, but they found no mains to be economically justified for replacement at the time of the study.

[4] To interpret what the studies concluded you must consult the original studies to assess the confidence in statistical estimates and model results.

[5] The coefficient of determination (R^2) measures the proportion of total variability of observed data that is explained by a model.

Kettler and Goulter (1985) evaluated regressions for the number of breaks versus diameter and time for cast iron and asbestos-concrete mains in Winnipeg, Canada. They showed strong inverse linear correlations between failures and diameter. Failures increased linearly with time with R^2 between .56 and .81. Soils in this area were known to destroy joint bolts in a few decades. Asbestos-concrete failures were dominated by circular cracks.

O'Day et. al. (1986) analyzed data from six utilities representing diverse geographic, demographic, and geologic conditions. They explained data needs and mains inventory and repair issues. Their major points dealt with standards for the identification of mains and repairs, compatibly with automated mapping, unique identifiers such as nodes and identification numbers for inventory elements, and cross-referencing of repair records to these elements. The study did not deal with whether the failure was on the pipe body or joint, etc., non-failure events (third party hit or street work), condition of pipe, information about the main (material, diameter, depth, etc), and situation at failure (water and air temperature, pressure, etc.).

The report by O'Day et. al. (1986) is a good reference to show the status as of 1986, when break prediction methods had just appeared. It summarized statistics on failure rates at utilities, types of main deterioration and contributing factors, ways of assessing main condition or risk of failure, economic methods for evaluating needed replacements or rehabilitation, needs for break and leak definitions on repair records, mains inventory and databases, leak detection, condition monitoring, and mains testing.

Individual pipe failure models were attempted in a 1980s research program sponsored by EPA at MIT (Andreou et. al. 1987a, 1987b; Marks et. al. 1987). They studied two data sets from large utilities in the Northeast and the same Cincinnati data as Clark et. al. (1982). Their goal was to assess the expected number of future break events for each individual pipe. They sought a method to analyze pipe records but not to present a universal formula because site specific studies would still be required. They defined and analyzed probability of a segment breaking at any point in time (failure rate per year).

They reported that three types of models were in use: aggregate models; regression type models, where expected number of breaks and/or time to break are predicted from environmental and pipe variables; and probability models where failure probabilities are estimated. All three approaches were "severely limited by data availability," meaning that they were not practical for use at that time.

The team found a dividing point at the third break and named the stages early-stage (slow breaking stage) and later-stage (fast breaking stage). Early-stage is distinguished by few breaks, long intervals (several years) between breaks, and decreasing time-between-breaks. Later-stage is characterized by more frequent breaks (more than one break per two years) and no pattern in time-between-breaks.

For early-stage failures, the team studied failure-time models with hazard-rate and survivor functions. Survivor functions predict probability of surviving beyond a future period and hazard-rate functions show probability of failure for a particular pipe during a specified period. After deriving the equations, they arrived at a basic equation with a baseline hazard function that represents the aging process and a stress dependent term. This model form is intuitively correct because, while the aging process proceeds, the pipe is exposed to additional threats.

In the proportional hazard model, given an estimate of the baseline hazard function, differences in deterioration versus time between segments are assumed to be proportional. It does not impose an overall functional form (linear, exponential, or other "parametric" model form) on the time dependence of the failure process. Ideally, it separately estimates an empirical baseline hazard function, and a time-independent function of the mains attributes.

The MIT study was the most substantive to that date, comprising a PhD dissertation (Andreou, 1986) and faculty research. It did not lead to a model for main failure prediction that utilities could adopt, but it did break ground in this area of research. The audience of the reports appears to be researchers not utility engineers. The study uses predominantly technical and statistical terms.

The study found the prevailing shape of the baseline function to be a "bathtub" function of age, decreasing early and rising again in the future. This does not allow for some segments not showing early problems nor for others beginning major problems at different future times. The model predicts that late breaks are random in time, which seems counter-intuitive as it seems that in later stages mains will exhibit increasing break rates.

It provided a mathematical form for failure equations and used pipe break data to show how risk-based and data-centered models could be formulated. The model exercise showed interesting things about the data sets. You could generate a histogram of relative hazard per foot, survivor functions for individual pipes, and change in hazard rate if a variable changes. You could also study the relative influence of factors such as internal pressure, land development, past break history, and year of installation, but the results were dependent on data validity and model accuracy. The large amount of statistical information could not be easily translated into models and the complexity of the models made them infeasible for use by most utilities.

The Cincinnati data had unusually high numbers of breaks (over 30 for some segments) and with uncertainty about whether break records included repairs, doubt remains on conclusions related to high break rate models. The Cincinnati sample was only 367 segments. Another limitation is analysis based almost entirely of mains greater than 6 inch diameter.

Regardless of its limitations, the MIT study reinforced prior findings that different models may be necessary for different stages of deterioration, showed ability of multiplicative factor model forms (as opposed to linear relationships) to represent failures in at least their middle stages, showed methods for modeling breaks as probability rather than as time-to-fail or counts of breaks, and evaluated the proportional hazards model form, which was new to water distribution research. The study set an example for statistical evaluation of main break data. It also showed methods of including segments with incomplete data, such as missing installation year or years of survival without failure. The data on these segments is important to determine relationships between differences in survival and differences in environmental factors.

Goodrich et. al. (1987) reported on time to first failure. They described a link-node based organization of databases for Cincinnati, OH (same data set as Clark et. al. (1982) and Marks et. al. (1987) and Kenosha, WI mains. Mains link (segment) data is available for each change in diameter, pipe intersection, tee, pressure control, and valve. They also used a proportional hazards model. Significant, positive linear relationships to break probability were shown for pressure, length, and the multiplicative combination of diameter, period-of-installation, and installer (utility or contractor). An inverse relationship was found to diameter. The major part of this paper described the preparation and management of the data for breaks prediction models and their application in mains maintenance. Each segment also has data on street and address range, service connections, and installation year. Historical repair records have been matched by address to the links in these systems.

Goodrich et. al. (1989) surveyed 12 utilities. They classified 14 mains and break data items in 5 levels of availability. The mains and repairs attributes were: diameter, material, age, soil type, pressure and pipe thickness, plus repair date and time and clarity of definition for the breaks records. Only 4 utilities computerized their data and only one coded a majority of the 14 attributes in the survey. For break records the only attributes that were readily available for most utilities were diameter and material.[6]

As reported in (Boynton 1989), the Des Moines Water Works had analyzed soils data and correlated it to breaks. The paper reported on soils data on a group of mains which had not yet broken as a control group. All mains analyzed were cast iron. Analyses were linear multivariate regressions of age on factors such as soil resistivity, pH, redox and sulfide measures. The results were mixed. Boynton (1989) also reported that beam breaks per month correlate to average monthly temperatures and hole and split breaks correlated to pumping volume.

Cullinane et. al. (1989), in an ASCE Task Committee Monograph on Risk and Reliability Analysis, compiled break rate statistics for several utilities and reviewed studies and utility experiences. The monograph also gives the relative break rates at approximately a dozen U.S. utilities. It devoted attention to the Clark et. al. (1982) study on Cincinnati, OH, New Haven, CN, and Kenosha, WI data.

Hosoi et. al. (1989) presented tables and plots of failure rates in Japan and North America versus diameter, type of material, joint and body/pipe, temperature, and city. With the exception of a few high rate cities in America, Japan failure rates are over twice as high, with areas other than Tokyo being the highest. Unfortunately inventories by material were not given, so it was not possible to separate out the dominating material in many of the leak rates given, but their data covers cast iron, ductile, AC, steel, and PVC. The ductile iron failure rates were very small. In Japan, PVC and asbestos cement pipes were shown to have higher failure rates than cast iron or steel in some cities. These two higher failure rate materials had lower diameters than the others, their failures varied inversely with diameter within this lower range, and their failures are predominantly at joints. Cast iron failures did not exhibit as much dependence on diameter as the other materials, especially over a range of 3 to 8 inches. None of the materials showed changes in failure rates down to about 41 degrees F, but cluster in the range of 40 degrees F for cast iron, and between 36 and 41 degrees F for all materials.

Mays (1990) demonstrated that there is an economically optimal point of time for action in the tradeoff between preventive and reactive maintenance for any infrastructure component.

Kulkarni and Reid (1992) reported a research project by the Gas Research Institute on methods for optimizing replacement versus repair in cast iron gas distribution. The methods are based on a framework for road network maintenance optimization and work at gas companies. The project developed cast iron main leak and break prediction models and life-cycle benefit/cost measures. The life cycle costs include risks to property of gas escaping and damaging property or injuring people.

The models, named CIMOS (Cast Iron Maintenance Optimization System), were field tested and validated at five gas companies. Software and methods were found to be useful and applicable to gas companies. This work is also described in a report for the Gas Research Institute (Kulkarni and Reid, 1992).

Repair data has been available at almost all gas companies for many years. Standard gas company (GIS) databases are driven by the computerization of maintenance functions such as work management, maintenance, and surveys.

---

[6] Note that the survey reported in this present project showed that the data situation had improved by 1998. It has further improved by the time of this report preparation (2006).

As with the Marks and Clark references, the study uses repairs as the definition of a failure. Later stages of this work have concluded that for most pipeline networks a single prediction function does not adequately describe failures well for mains that are just beginning to leak and those with previous leak histories.

Kulkarni et. al. (1990) described an optimization system for decisions between repairs, rehabilitation, and inspection/testing for transcontinental gas pipelines. Since failures are infrequent but consequences are large, the inspection alternatives are important. The transmission model, named PIMOS, is based on a GIS database of pipeline characteristics, maintenance records, demographic, and environmental data along the lines. The prediction models on the high pressure steel transmission lines are simpler and more generic than for gas main failures. Decision trees of maintenance and rehabilitation options and methods were developed.

An AwwaRF report by Kirmeyer et. al. (1994) included information on modes of pipe failure from an extensive survey. Other surveys and reports described in Chapter 3 show different results, depending on the utilities that participate.

Kane and Gough (1996) reported on a prioritization system for mains replacements at Severn Trent Ltd., the company that privatized the 25,000 miles of distribution mains in the Birmingham, England area. Priorities for District Meter Areas (DMAs) were based on water quality, aesthetics (discoloration, odor), service interruption, pressure loss, structural condition (age and history of bursts), maintenance costs, and leakage (damage, cost, supply). Each of the 1800 DMAs may cover a population of several thousand people and averages 14 miles of mains.

Water quality, aesthetics, some leakage factors, and structural condition used in this system are based on monitoring of these factors, not expressed in cost terms or calculated in terms of breaks. They use experience and opinion to set weights on the perceived severity of the factors within the major categories, based in part on customer surveys and maintenance cost history.

They used a hierarchical weighting system for the performance factors in prioritization, not based on predicted breaks. The consequences model has subjective weights for policy or sensitivity analysis. It records main age categories and histories of burst and leaks, as well as maintenance costs.

Elnaboulsi and Alexandre (1997) derived a single functional form of the multi-stage proportional hazards models described by Marks et. al. (1987). This could ease computational difficulty of these models. They did not report comparisons with data sets but they collected economic consequences of breaks, and used them with their break model derivation to determine the optimum year to replace segments. They defined and collected data on direct and indirect economic consequences of breaks and used them to determine the optimum year to replace mains in the Paris and Strasbourg, France water systems. The considered replacement costs, consequence costs, and discount rates. Direct cost data used were for repairs, extra pumping, and water lost. Indirect costs used were user damages and loss of profits. Other possible consequence costs were discussed, to include: traffic delays, street and utility damage, hospital loss of water, water customer inconvenience, fire protection, electric and computer installation cooling, work stoppage, flooding, sensitive agricultural operations, and water contamination.

Lim (1997) reported on a mains evaluation system "PIPES" in the City of Seattle. PIPES had consequence factors in its vulnerability and criticality submodels. These were linked through its GIS database to the locations of predicted breaks. The vulnerability sub-model considers both vulnerability of the pipe and of the public. The criticality model considers land use near the main, such as hospitals, dialysis centers, schools, major water users, community centers, and flood-prone areas. The system shows implementation of consequence factors via a GIS database to prioritize and evaluate problem mains,

Lyonnaise des Eaux's (Madioc et. al. 1996) PREDIKASS model is a tool for failure prediction and condition assessment. PREDIKASS was reported to be require 6 to 10 years of mains data. This was the only data from Europe to be provided to the team, and was not verified at the time.

Thames Water Utilities, Ltd. (1998) reported "Utilnets" as a hydraulic model to estimate burst flows with elevation models for route and fate of water and priority data for consequence analysis. The model applies to trunk mains only.

Deb et. al. (2002) reviewed utility programs to collect data and assess reasons for pipe failure. They described the 1980s efforts of the Philadelphia Water Department to conduct visual, mechanical, and soil tests on pipe failures. Their review of a Severn Trent Water study showed that most breaks occur in CIP and that corrosive soils and expansive soils lead to more breaks. Cold weather causes breaks and most breaks are circumferential. The Severn Trent study found that breaks decrease after cleaning and lining, a finding that is at odds with the modeling efforts of Andreou et. al. (1987a).

They also proposed a method to prioritize individual pipes. They incorporated two models of pipe behavior: a pipe load model and a pipe deterioration model. These corresponded to the terms in the equation of Andreou et. al. (1987a). The pipe deterioration model, corresponding to Andreou's baseline hazard function, includes information from pipe characteristics and soil samples. Some soil properties, such as pH, moisture, resistivity and aeration, combined with pipe properties and water quality, may lead to deterioration over time. The pipe load model takes into account other soil properties that influence loading such as moisture, bulk density, grain size and expansiveness. Other loading factors such as frost, earth and truck load are also considered, along with internal pressures like water hammer and thermal changes. These loads are included in the model and used to determine stresses on the pipe. The two models, along with break history, climate data and safety factors, predict the expected condition of a pipe and were intended for use to prioritize main renewal programs. However, lack of validation hinders application of the models.

A number of utilities have developed models for their own prioritization and capital planning. The methodology of the Los Angeles Department of Water and Power (LADWP) was presented by Mavrakis (2003). They use an A-F rating scale, with age-related deterioration scores. For pipelines, the maximum service life was assumed to vary from 70 to 210 years depending on pipe material and soil corrosion index. Three soil types are considered: severe corrosion (30%), moderate (50%), and mild (20%) corrosion. Pipeline grades were determined separately for steel and cast iron/ductile iron. Trunk lines were evaluated, and their lengths were taken from the trunk line hydraulic model. Some 743 miles of trunk line (16" and larger) were evaluated on an overall basis at "B", and with a unit cost of $13 per inch of diameter per foot. Of the lines, some 70 miles were rated D, F, or critical. The present value of replacement cost of all trunk lines was $1.67 billion. This analysis yielded an estimate of $54.5 million needed now for replacement of critical mains and $10.1 million needed annually to keep up with critical replacements over the next 20 years. Considering all assets, including tanks, reservoirs, treatment plants, and trunk lines, a cash flow replacement schedule similar to the Nessie curve was produced. Total average replacement needs are $16.2 million per year.

Columbus, Georgia has a large inventory of small cast iron and galvanized steel pipe (3" and smaller) that need renewal. The inventory is estimated at 167 miles, with a replacement cost of $52.4 million. They undertook a GIS-based study to identify lines for replacement, and used several information systems to get information on breaks, low pressure complaints, and water quality complaints, to go along with age and other data that is used in their prioritization scheme. The scheme is based on a voting system, with priorities in order being customer complaints, water quality, maintenance, fire protection, materials, age, and pressure. They also decided to replace some of the service connections as they replaced small lines. The winning bid was $58.67 per linear foot in 2000 dollars. They also compared their capital improvement program to the Nessie model, which requires assignment of service lives to categories of pipe. They compiled age data from different sources, including employee interviews, but the data varied and indicated some uncertainty. Lessons were that GIS is a useful tool for this kind of study and for asset management, age dating is useful, access to data bases is required, and just getting the pipe data represents good progress (Horn 2003).

The City of Edmonton, Alberta began a reactive prioritization program in 1985 (Seargeant, 2003). In 1985, there were approximately 1,600 main breaks in Edmonton. The number was reduced to 468 in 2000. This was accomplished through a program in which a surcharge was added to customer bills dedicated solely to replacing main breaks. The surcharge equated to about $3 to $4 per customer. Seargeant mentioned the application of a proactive program in some cases, using hydroscope results, a hydraulic model, and the coordination with other utilities. However, the noted success was from a reactive program. Replacements were based on service level only in which all pipes with a break frequency of greater than 5 breaks/km/year, based on a five year average, were replaced.

Hope (2003) developed an approach for the prioritization of water main replacement based on assigning a priority action number (PAN) to each main based on agreed service level goals. Several service level criteria were established including pipe age, leak history, water quality, criticality, hydraulic performance, corrosion, material, system pressure, and location. Once PANs are determined for each category, categories are weighted according to importance. After the numbers are assigned, a prioritization model called OWASA is created. In it, a variety of data sets are integrated into one geo–database, such as operational work orders, a sewer system hydrologic model, customer service billing, and a water system hydraulic model. From this integrated database, a water main prioritization model can be developed, along with an overall utility capital improvement plan. With all of this data integrated, sections of pipe are designated as highest priority for replacement.

Canada's National Research Council has modeled risk of individual pipes (Kleiner, et. al., 2004). They found that "distress indicators" are required and are inherently imprecise and involve subjective judgment. Distress indicators can include variables such as age, history of leaks, frost load, and other threats. They thought consequences of failure for buried pipes are difficult to quantify because of lack of data. They used a "fuzzy rule-based, non-homogenous Markov process" to create a deterioration model. The application of fuzzy set theory appeals to researchers dealing with scattered data because it offers them a way to consider uncertainty within uncertainty. To train the model one needs to know or assume the condition immediately after installation and to have one condition data point at age t. The authors add risk to their model by including assumptions about consequences. Without data and verification it is impossible to confirm the validity of this approach, however.

In addition to these reports, Rogers (2006) reviewed several others on main failures. For example, Agbenowosi (2000) developed a mechanistic model to predict pipe failures, but it is unknown whether he tested the model on any data. Kleiner and Rajani (2001) published a review of statistical methods to model deterioration of water mains. Loganathan, Sherali, Park, and Subramanian (2002) presented a method to truncate data sets, but Rogers did not indicate if they had used data sets. Pelletier, Mailhot, and Villeneuve (2003) presented three case studies of water main breaks. Vanrenterghem (2003) studied the New York City water system.

Meanwhile, additional studies and datasets for main failures and models continue to be reported. For example, SEAMS (2002) prepared a model for infrastructure decision support for U.K. utilities.[7] The model appears to be based on pooled utility data for a single utility, but to represent another approach that can be considered by utilities for their capital improvement planning.

AGGREGATED MODELS FOR SYSTEMWIDE PLANNING

Aggregated pipe models may offer useful methods to compile information about network condition and need for renewal, but they do not give information on condition of single pipe segments. The KANEW model, whose name derives from a European model ("Karlsruhe New"), provides a method and software for a "predictive distribution system condition assessment model" (Deb, et. al., 1998). It can be used to quantify system costs but not to prioritize repair, rehabilitation, and replacement for individual pipes.

KANEW works by importing data about the network. Data are differentiated according to year of installation or rehabilitation and with respect to aging behavior. The most important criteria are age, material, diameter, and bedding quality. Then, survival functions are determined to predict life expectancy. These are based on records of past failures and on expert opinion about survival of newer pipes. Taking uncertainty into account, the model offers pessimistic and optimistic scenarios for pipe survival on an aggregated basis.

The advantage of KANEW is to allow the utility to use a forecasting system and to budget over the long term. Limitations of KANEW are that it is general, not community–specific, and is based on limited variables. Also, the KANEW software can only be run on older PC platforms (Windows 98).

The "Nessie" model also considers aggregated systems (Cromwell, et. al., 2001). Utilities can use "decay curves" derived from statistical analysis of break data to forecast repair, rehabilitation, and replacement needs for classes of assets. This can be projected to create the Nessie Curve of future investment needs. The Nessie curve shows waves of increasing investment needs for utilities. It aims to predict when these waves will occur and how they can be avoided. Creating the model requires a great deal of data on pipe characteristics but it does not tell you which pipe to replace.

ONGOING AwwaRF RESEARCH

AwwaRF is continuing with research on risk factors for pipe performance. In 2006, it announced availability of three reports related to this project:

---

[7] The SEAMS report was described to the AwwaRF Research Advisory Committee on Infrastructure Reliability at its July, 2006 meeting in Denver. The report and its companion documents and software have not been reviewed in detail, but the reference is provided for future studies.

- Long-Term Performance Prediction for PVC Pipes (#91092F)
- Potential Techniques for the Assessment of Joints in Water Distribution Pipelines (#91126)
- Risk Management of Large-Diameter Water Transmission Mains (#91087)

The report on long-term performance prediction for PVC pipes tracks some of the research in this present report, and contains a similar recommendation. It studied manufacturing and installation practices for PVC and noted experiences with failures, including the U.K. "nationally agreed database of water main failures." An important recommendation, quoted from (Burn, et. al. 2005) is: "Because of variations in the methods of reporting data and the difficulties in cross-correlation, it is recommended that an effective method of recording and organizing pipe asset and failure data be used, to capture this variation. Until data can be represented in this manner, care must be taken in interpreting and comparing failure rates."

Risk management of large diameter water transmission mains was studied by Kleiner, Rajani, and Sadiq (2005). Their conclusions about data are the same as reported in this study: data are scarce and utilities require some way to use judgment along with data to make renewal decisions. The authors propose a method based on "maximum risk tolerance" to be used in conjunction with inspection data. Large diameter mains can often be inspected more readily than smaller mains.

The report on potential techniques for the assessment of joints in water distribution pipelines was not available as of the preparation of this report.

CHAPTER 3
MAINS BREAK PREVENTION SURVEY

ABOUT THE ORIGINAL PROJECT SURVEY

The original project included a survey of utilities to study data needs and availability for main breaks control. The survey was conducted in 1998 so the findings will be some 8 to 9 years old by the time this report appears. The results are presented in chronological perspective with other surveys.

The survey was conducted in cooperation with AwwaRF's project "Prioritizing Water Main Replacement and Rehabilitation," which yielded the report in (Deb. et. al., 2002). Its survey instrument was 5 pages with some 110 potential answers distributed among 9 questions (see Appendix). The principal questions were on data availability, quality, and definition or documentation. To encourage utilities to respond, they were called in advance and sent a research summary to stress benefits of participation. A follow-up call was made within four weeks if necessary. By coordinating with AwwarF project "Prioritizing Water Main Replacement and Rehabilitation" and making a random selection from some 1,000 utilities (from the 1996 survey in AWWA's Water://Stats) some 65 surveys were sent to utilities with an average size of about 100,000 population served. An additional 10 were sent to international utilities. Some 47 completed surveys were returned within 6 weeks for a response rate of 63%.

FINDINGS

Major findings included (see the Appendix for a summary of findings and of the survey questions):

Inventory and break rates

- The average length of their distribution systems was 1,938 miles (ranging from 418 to 7,040). The average number of their service connections was 165,000 (ranging from 47,000 to 800,000).
- Responses showed that the average mains inventory of this sample of utilities (serving 100,000+ population) is comprised of 52% cast iron, 21% ductile iron, 12% plastic and related, 10% asbestos-concrete, and 6% steel or galvanized.
- Average annual break rate (for all materials) was .25 per mile.
- Cast iron break rates were significantly higher than the other materials.

Expected Value of Research to Utilities

- Some 72% of respondents said the research appeared to serve their needs and the most important reason was ability to justify replacements.
- Of those who said the research did not seem useful, the reasons most cited were data, cost, and skill limitations.

Priority-setting

- The utilities rated breaks as more important than other factors (such as capacity, water quality, fire protection, transmission line effects and coordination with public works) in prioritizing replacements.
- The utilities gave mixed answers to the question about how they selected mains for replacement.
- Assessing mains life was a clear issue but assessing costs was the least problem on average.
- On average utilities rated breaks as 63% more important than other considerations (such as capacity, water quality, fire protection, transmission line effects and coordination with public works). If utilities gave low weight to breaks, they had below average break rates.
- Some 80% of utilities preferred that research focus on segments of mains rather than categories of mains.

Current practices in prioritization

- Some 13% of utilities reported a points system to target replacements. Some 24% of utilities used a categorical method for selecting replacements (for example, mains over 60 years old and under 5-inch are selected).
- Some 34% of utilities determined replacements by expert opinion or committee.
- Few have tried a breaks prediction method from the literature and some developed their own method.
- Reasons for lack of implementation of models focused on data and staff limitations, as well as labor costs, knowledge, and availability of alternate methods.

Data availability

- Eleven factors (such as address, date, break versus leak, type of break, etc.) were available from most utilities and about half of these reported they had the data in computer databases (later results showed that these were optimistic estimates of data availability, for example, soils data and temperature at failure is rarely recorded).
- Data on attributes of mains not failing is also required but not emphasized.
- Data on all mains is available or being compiled in mapping software at some utilities and others have mains data available only on maps. The attributes available in almost all mapping software systems are material, diameter, street name, geographic coordinates, and length. Year installed was available in 78% of those with mapping software. Some 87% of utilities may have data in automated mapping systems within a few years.
- Availability of consequence data was very limited.
- On average databases covered 14.6 years of breaks information.
- Utilities do not all record whether a "repair" is a planned action, such as streets coordination, opportunity work, or a proactive action. It is necessary to separate "repairs" which were actually due to failures.

Consequences

- Availability of consequence data by utilities was very limited.
- Break consequence that utilities listed as most important was outage hours (versus claims, repair costs, etc.)

COMPARING THIS SURVEY TO OTHER SURVEYS

Results from this survey can be compared with others that have been taken during the last ten to twenty years. For example, AWWA's water industry surveys began before 1900, and the information they provide can serve as a current source of information about water distribution infrastructure.

Surveys were conducted every five years from 1945 through 1970, then surveys were conducted more frequently (Keeley, 2003). The survey report was renamed WATER://STATS in 1996, and since then, surveys focus on specific subjects, such as finance (1999 survey of 672 utilities), distribution systems (2002 survey of 337 utilities) and rates (2004 survey).

Prior to about 1980, the surveys focused on water production and rates. Management attitudes have changed, requiring more data now. Concern about high levels of deferred maintenance and other concerns about distribution systems increased after 1985, especially after the SDWA Amendments of 1986.

A 1985 survey was reported at AWWA's 1985 Distribution System Symposium (O'Day et. al., 1986). The survey showed that actions differ among small, medium, and large utilities and by comparing its results to later research, you see that survey questions and responses often create uncertainty in the findings. For example, Table 3.1 taken from (O'Day et. al., 1986) suggests that most medium and large utilities have reports of main breaks, methods to determine replacement needs and budgets for replacement. In the light of later research, this result would seem to be too optimistic.

Table 3.1
Utility practices in replacement planning

	Small <5 mgd	Medium 5-50 mgd	Large > 50 mgd
Method to determine replacement need	63	77	100
Reports of main breaks	27	66	80
Budget and planning for replacement	48	67	90

Source: Data from O'Day et al. 1986

Kirmeyer et. al. (1994) surveyed twenty utilities in 1992. They also summarized AwwaRF's three expert workshops on distribution systems, held in 1990, 1991, and 1992. Statistics from their survey are comparable with the latest findings (Table 3.2).

Table 3.2
Inventory of distribution systems

Estimated length of distribution piping	880,000 miles
Estimated number of pipe breaks	237,600 breaks per year
Type of existing piping	48% cast iron, 19.2% ductile iron, 15.1 % AC

Source: Data from Kirmeyer et al. 1994

In addition to these statistics, Kirmeyer et. al. (1994) reported on causes of main breaks (Table 3.3):

Table 3.3
Causes of main breaks

Causes of main breaks	Percent of utilities reporting
Materials/deterioration	55
Weak joints	35
Earth movement or settling	30
Freezing	30
Internal corrosion	25
Corrosive soils	25
Construction or utility digging	25
Stray DC current	20
Seasonal changes in water temperature	15
Heavy traffic load	10
Tidal influences	5
Changes in system pressure	5
Water hammer	5
Air entrapment	5

Source: Kirmeyer et al. 1994

They also reported criteria for pipe replacement, as shown in Table 3.4 (Kirmeyer et. al., 1994):

Table 3.4
Criteria for pipe replacement

Criteria for pipe replacement	Percent reporting
Number of leaks or breaks	75
Age of pipe	45
Low flow	40
Condition or type of material	30
Size changes required	30
Water quality	15
Soil condition	15
Location	10
Street construction work	10
Elimination of dead ends	5
Amount of damage by leaks/breaks	5

Source: Kirmeyer et al. 1994

Deb et. al. (2000) surveyed 37 utilities about O&M, including European utilities. Some 60% reported procedures for main breaks and 70% had maintenance history databases.

AwwaRF Project "Prioritizing Water Main Replacement and Rehabilitation" cooperated with the original project in the survey (Deb, et. al., 2002). This survey found similar results: break rates of .21 to .25 breaks per mile per year; European utilities have higher break rates (on order of .50 per mile per year); and European utilities have more organized data programs and remedial programs for controlling breaks than North American utilities. This project imported water main break databases from six utilities.

In their survey, 70% of North American and 85% of European utilities indicated they had computerized records of main failures (however, this does not reveal how well the data is managed or maintained). They believe that useful applications of data can be made, even with limited data. They recommended a comprehensive program of main break data collection consisting of field, office, testing, and cost data. Recommendations were for all utilities to collect and maintain water main break data, and guidelines were presented. The guidelines show 45 data items to be collected for a main break. This illustrates the complexity of the problem.

AWWA's 2002 Distribution Survey (AWWA, 2003) includes 337 utilities in the United States and Canada. The survey covers pipe materials, valves, fire hydrants, finished water storage facilities, corrosion control, pumping capacity, metering, customer service lines, water auditing, leakage management and infrastructure needs. The 337 utilities that were surveyed serve 59,389,902 in population, and have 14,339,261 customer service lines, and 146, 435 wholesale connections.

This is a comparison of 2002 AWWA data with 1992 data (Table 3.5):

Table 3.5
Comparison of distribution system data

Item	**1992 Survey Kirmeyer et. al. (1994)**	**2002 Survey AWWA**
Miles of pipe (US total)	880,000	980,000
Expansion, miles/year	13,200	5,100
Replacement, miles/year	4,400	3,590

The 2002 Distribution System Survey included some main break information. By removing obvious errors and outliers, the average main breaks were 0.21 per mile per year and the average reported cost per main break was $1,320. The questions were simple and did not explain to the utilities how to define a break or which cost items should be included. Nevertheless, the data for break frequency is in the same zone as other surveys.

CONCLUSIONS ABOUT PROJECT SURVEY

The information collected from the original project's survey seems consistent with other survey data collected since the mid-1980s. The difficulty in collecting and interpreting this data is the ambiguity in questions and their responses. For example, a question such as "do you have a procedure for planning main replacements" may get a "yes" answer, but it is difficult to know how serious the procedure is.

CHAPTER 4
LESSONS FROM MODEL DEVELOPMENT AND TESTS

PURPOSE OF CHAPTER

This chapter summarizes lessons learned during the original project about modeling of main breaks. It can be useful to utilities seeking detailed technical information and to researchers seeking to develop future models. The chapter is an edited version of material from the draft report for the original project (URS, 2002).

The chapter is technical in nature and can be skipped by readers who are more interested in management applications of the model, and less interested in the technical details of modeling.

The original project sought to develop software to apply break prediction methods and consequence data to develop tools to prioritize replacement projects. Users of the software were intended to be utilities with complete mains datasets and those with only break records.

ASSUMPTIONS FOR MODEL DEVELOPMENT

Assumptions for model development were:

- The reason for breaks prediction is to support decisions about replacement.
- A model should use segments as the unit of analysis and replacement.
- It is feasible to model future behavior, given recent performance and the attributes of a water main.
- A main does not age in isolation but is modified by repairs and maintenance.
- With data covering only a small part of the life of a main, you cannot know if data is for a first break or another break in a main's life.
- A repair would slow the segment's natural aging. Thus, to be useful the prediction model should represent a segment's behavior after it is repaired.
- A prediction model should consider different periods of break history and main repairs as well as cumulative breaks.
- If non-breaking data is unavailable, classes of mains should be used to preserve the mix of breaks history and segment attributes
- It is desirable to predict both first breaks and later breaks.
- The availability of 8 or more years of breaks history is required to support models.

DATA COLLECTION AND PREPARATION FOR THE MODEL

Characteristics of mains

The project's sample of breaks over 7 to 12 years for 5 utilities showed an average age of 60 years for cast iron mains with at least one break. The worst 20% had an average age of 64 years. Of the utilities with steel mains, the segments with at least one break in the last 9 years averaged 45 years old. Those being replaced averaged 60 years. Replacements were being selected from the worst one third of those breaking. The range of replacement ages for both materials had a standard deviation of 14 years.

Experience in collecting data

The steps to obtain and document a utility's breaks database were: 1) specification of required variables, 2) assessment of data availability, 3) specification and request for data, and 4) follow-up to improve data collection.

Utilities usually had less data than they thought. At least 6 years of breaks records were said to be available at these utilities, and many more years at some. However, follow-ups indicated that only three utilities had databases sufficient for model estimation. All had 6 to 16 years of coded data with material, diameter, installed year and break type. Most had one or two additional breaks attributes (pressure, joint type, soil type, etc.)

Four utilities had inventory attributes at a segment or installed reach level (but two of these had no basis for matching their break records to their segments). Except for material, diameter and installed year, inventory attributes were not universally available, missing for many records, or weren't available in their breaks records.

Only one of the 9 utilities had all mains and breaks data in a Geographic Information System (GIS). That utility supplied its mains and matched breaks data by simply querying their GIS system. Their segments were installed units with most being longer than practical for analysis and prevention.

Four utilities had no attribute data on non-breaking segments, or were unable to match breaks records to inventory segments. This indicated that many other utilities would have these limitations and challenged the research objective of predictions covering mains that had not yet broken. Thus, a method was developed to describe mains attribute distributions based on existing aggregate data or expert opinion estimates.

Data to support modeling was often incomplete and not reliable, requiring much quality control, cleaning and transformation. Given the quantity of data, this requires automated procedures, which are expensive.

Of 14 data sets used to model failures in the literature or collected by this research, only 3 covered more than 12 years of breaks and most cover 10 or fewer years. A data set for Bordeaux, France is said to cover 42 years of breaks, with 33 years available for modeling (Madioc et. al. 1996). A data set for Cincinnati covered 40 years of breaks but modeling was on only 2% of the mains of large diameters (Marks et. al. 1987). Clark et. al. (1982) studied 16% of Cincinnati's mains.

How to handle data

Processing is required to prepare data for a prediction model. Classes of mains and breaks with different failure mechanisms should be stratified (e.g. CIP, AC, steel, lined/unlined, break types). Making street names consistent is required. Most utilities lack segment IDs matched to breaks and street names. Break history variables are required (failures in 3, 6, and/or 10 year intervals for each year). The model attempt required the processing of approximately 50,000 breaks records from 6 utilities.

Quality Assurance and Control (QA/QC) involves:

- Basic QA (clear definitions, missing data, completeness, and consistency)
- Data Cleaning (inconsistent or misspelled street names; code set errors)
- Data Transformations (proxy or correlation rules for missing data)

- Reformatting of fields

Data QA is laborious but is eased with analytic tools. These search for certain codes, calculate statistics on large data files, and perform other tasks. Basic tools are available in spreadsheets and more power is in statistical analysis packages. Much of the effort in the project was devoted to algorithms for data quality control.

Street name consistency is a special case and requires special effort. Desirable data attributes that were checked for and processed:

- Distinctions between major breaks (circumference and blow outs versus holes and split/longitudinal)
- Eight or more years of complete breaks data
- Identifying and obtaining mains attributes by segment units
- If mains segment data is available, the ability to match breaks records with segments
- Expert opinion information on the distributions of inventory attributes
- Utility data quality assurance and control

Data from utilities

Data collected from utilities is shown in Table 4.1, which compares statistics for inventories and breaks records from the 6 contributing utilities. The statistics on break rates are also compared with the breaks survey and with the AWWA Water Stats database. Also shown are the mileages of the three materials with the highest break rates per mile, the ones considered in this research. The last column shows the number of years of data which the models had to work with at each utility.

Table 4.1
Comparison of collected inventory data and break rates[1]

Utility[2]	Miles	Yrs data	Brks/yr	% of material/% of breaks			
				CIP	AC	DI	Steel
A	1,300	12+	269	.95/.21	0/0	0/0	0/0
B	2,530	6	482	.13/.235	.17/.138	.1/.0114	.54/.24
C	7,100	10	1,700	.70/.19	.09/.061	.06/0	.15/.607
D	1,474	26	208	.44/.253	.003/0	.38/.049	.10/.095
E	3,759	7	750	.40/.35	.31/.104	0/0	.03/.045
F	3,200	27	990	.56/.40	.02/0	.28/0	0/0[6]
Survey[3]	1,817	15	421	.47/[5]	.12/	.24/	.07/
Water://Stats[4]	1,323	5	380	.36/	.17/	.24/	.04/

Notes to the table.

1. Data in table has not all been checked or validated. It is taken from a draft report.
2. Utilities were identified in draft report. They are large public and private utilities.
3. Survey of utilities with greater than 100,000 population served.
4. Water://Stats was AWWA's 1996 survey.
5. The break rate averages were .242 in the survey and .287 in Water://Stats. These were not broken out by pipe material.
6. A zero indicates a small or negligible quantity.

Approximations of main attributes

In the project, a method to derive approximated main attributes was developed to incorporate data from both breaking and non-breaking mains. It was named a hybrid "cross-classified attribute record" (CCAR).

Data classes worked well to represent segments of pipe by diameter, material and year installed. Data on year installed can represent the continuous effect of age and the discrete effect of a decade of poor installation or material. The continuous aging effects can be approximated by classes such as high, medium and low. The discrete effects may take 5 classes. An example of a class for installed year on CIP might be years of spun cast pipe. For steel mains a year installed class might be the post WWII years when salvage iron pipe was being installed.

Correlations of attributes can be represented without individual segment data. This question arises if you extend a single attribute classification to cross-classifications of two or more attributes. For example, if you have 5 diameter classes and 6 installed year classes, you have 30 combinations. The percentage of mains by diameter will not be the same across installed year classes. You can compute percentage of mains of each from utility records. The correlations of these two attributes and with joint type were tested. Correlations were rarely as high as 20% between any of the attributes. The team concluded that only small errors would be made by assuming independence between key variables. Thus the percentage of mains in each cross-classification cell can be calculated with small error as the product of the percentages in their independent classes.

The researchers concluded that combined attributes of non-breaking mains and breaks records can be used in modeling and will not produce significant errors in the coefficients of the prediction variables. The less significant non-breaking mains attribute coefficients can have moderate errors with this method but only if they are highly correlated or if they constitute a large percent of the variability of the predictor variables.

The non-breaking main data is combined with break records by treating each cross-classified cell (such as 8 inch diameter installed between 1933 and 1945) as one record with a weight equal to the length of mains in that cell divided by the average length of a breaking segment.

Calculating percentages of non-breaking mains by attribute class given the total mains class percentages and the break records is accomplished by: matching break records to the attribute cross classified cells they would fall in, summing distinct number of segment locations represented by the breaks in each cell, and subtracting the sum of those segment feet from the feet of all mains in that cell.

A utility would supply mains attribute class data to a prediction model system. There are three levels of availability of mains attribute data: individual segment data that has not been matched to break records; data on the distribution of each attribute over all mains of a material (aggregate not segment data); and expert opinion estimates of the percentages of mains miles in key classes by material.

Utilities can estimate the percentages of mains mileage in each class of each key attribute, usually diameter and installed year. These percentages by attribute must be calculated or estimated for each material of interest. Calculations by this method were necessary on data from 4 of the 6 utilities. They produced the cross-classified non-breaking mains data for model estimation and algorithms for computing the non-breaking inventory statistics for the prediction software.

Cost data

Cost data were generalized for the model effort. A list of potential breaks consequence costs were sent to the six utilities contributing breaks data to the research, plus four additional utilities who agreed to supply them. Potential costs were ordered by their expected importance or magnitude, as shown in this list:

- Routine repair costs (planned)
- Emergency, large repair costs (unplanned), pavement restoration
- Landscape restoration, traffic control
- Sensitive use claims: large industrial, hi-tech, medical, etc.
- Property damage claims
- Lost water revenue
- Water quality, regulatory fines, lost confidence and goodwill
- Extra pumping
- Fire protection risk

In addition to the direct costs of repairing breaks, the consequence candidates included costs associated with the repairs such as paving restoration, and indirect factors such as claims and risk potential. All the costs related to breaks are to be measured in dollars. The indirect costs are dollars the utility would be willing to spend to avoid the factor. Nearly 60% of the cost items that were expected to be in the "important" category in the table were reported to be available at the utilities. A few of the additional cost items, initially judged as desirable, were also said to be available. However, within two months of the initial request, only one utility had supplied consequence costs. Most of the costs in the important category are necessary to effectively prioritize mains segments for replacements.

Some utilities claimed they maintain formal break costs accounts. In some cases these accounts are subdivided or flagged by costs of labor, materials, pavements, landscaping, etc. The accounts did not identify the costs associated with individual breaks but usually record dates or the year.

Given annual break statistics, it is possible to estimate at least average costs per break. A few utilities have separate accounts of major claims or fines. However, the utilities contacted indicated uncertainty in their ability to separate these costs or track them to main breaks only.

Only 5 of the 10 utilities that said they could supply breaks costs could actually supply adequate data. Only the utilities that contributed breaks data could supply breaks costs. Many said that they had only total or ill-defined maintenance cost accounts, without confidence that they applied only to breaks. Few said they could separate the cost components for different mains type, size or claims, much less by claims. The data still needed clarification to distinguish cost components such as pavement repairs and schedulable versus emergency or large break costs. A minority of the utilities supplying costs could identify claims, water quality or landscaping/erosion costs. The average break consequence costs that were submitted by the 5 utilities are shown in Table 4.2[8]:

[8] Note that these costs diverge from the break costs reported in AWWA's 2002 survey (See Chapter 3). The differences are probably due to lack of consistency in definitions and to utility accounting for break costs.

Table 4.2
Cost data on main breaks

Break cost item[9]	Average $/break, year 2000 costs
Ordinary repairs (schedulable)	$4,606
Emergency repairs (unplanned)	$7,432
Pavement restoration	543
Claims/property damage	248

Since the costs are from only 5 utilities they do not represent averages for the water community. However, the total per break costs ranged from $4,366 to $7,032, with a standard deviation of 18%, so there is good agreement for these 5 utilities. The characteristics of these utilities are shown in Table 4.3.

Table 4.3
Characteristics of case study utilities

Utility	Ownership	Miles	Annual breaks	Most miles in	% major material
California 1	Private	2,350	482	Steel	54
California 2	Municipal	7,100	1,726	CIP*	45
California 3	Municipal	3,760	857	CIP	43
Kentucky	Private	3,160	450	CIP	31
Canada	Municipal	1,473	305	CIP	52

* Cast iron pipe (CIP) was unlined

In addition to the unit costs of breaks above, the prioritization of mains requires replacement costs. To account for different length and diameter mains, costs should be per foot by diameter. In prioritization models for gas mains, up to 7 diameter categories have been considered.

MODEL DEVELOPMENT

This section describes development, parameter estimation, and testing of the main break prediction models. The models also merge break predictions and costs to create a decision support model for replacement.

Concepts

The information needed from the model are the dependent variables, and the input data are the independent variables. Also of interest are how the model works, and how it is verified and validated. The decision scenarios for which models are run involve changing values of the independent variables.

When model output is compared to observed data, it is called verification. When you then test this verified model on an independent set of data it is called validation. These steps

9 In addition to these cost items, the draft report included landscape restoration at $29 and fines and other costs at $4. These are so small that they are not included in the table.

follow the scientific method where something is observed and a theory is developed (verified model). Then the theory is tested by independent data (validation).

Past research showed that an individual attribute is not usually the best predictor of breaks. The mix of the worst values of attributes on a segment and break history indicate the highest probability of a break. Hence, in modeling it is most important to characterize the worst combination of key attributes that correlates to observed breaks.

Dependent variables

Ideally, we would like to know the main's breaking behavior in terms of its attributes, environment and break and repair history. Breaking behavior of a main has been defined in several ways, from the probability of having a certain number of breaks in the next year to the time between breaks. Characterizing these questions leads to identification of dependent variables.

Data limits

In practice main records are rarely available for as much as 12 years, a small minority of the lifetime of a water main modeled in those studies. Fortunately, long term models are not needed to justify replacements. For mains beginning to break, only short term predictions are necessary to justify replacements. Where longer term break or repair records are available, there are also serious questions of the consistency of the definitions of the data and the uniformity of maintenance practices over decades of use.

Questions to answer by modeling

During model development the following questions should be answered:

- What break types or severities are necessary to consider?
- For what upcoming period should break predictions be made?
- What mathematical functions fit the data?
- Which of the available mains and breaks attributes are significant predictors?
- Are individual segment level predictions feasible?
- Are different periods of break history variables significantly distinct?
- Are separate models required for different mains materials?
- Are model parameters transferable or unique to a utility?

Past models and promising methods

The past research identified six model concepts, and three were studied in the original project.

Main age at failure. The literature showed that linear or exponential functions of independent predictor variables can predict age at failure, at least in a theoretical sense. However, long term data is not available from most utilities. Only one of the 8 utilities providing data had more than 11 years of coded break data. Also, the models can predict

unrealistically low times to failure if they are based only on the mains that failed.
Total failures over time. This model type requires long term records, which are seldom available. It would use only main age and pipe attributes (e.g. diameter, pressure, environmental factors, etc.). If precursor events are not used, the model type would miss information that could improve prediction.
Breaks as functions of single causative factors. Single factor predictions can be used as first steps of analyses (break rates as functions of past breaks, decreasing diameters, etc.). These are popular and simple, but only explain a limited part of breaks behavior. These usually do not apply to conditions other than those from which they were derived.
Break probability in the next time period. These can be linear or non-linear expressing the probability of one or more breaks. Examples are the proportional hazards models by MIT and various exponential functions and linear combinations of main and break history attributes as in the French models.
Aggregate data. This is a data management tool rather than a model. It can accommodate attributes for aggregates of segments.

Data limitations

The importance of using data for non-breaking as well as breaking mains was explained earlier. Biases could be large if most main segments had not broken. Predictions will also be biased if they are based on aggregated data which was estimated from individual main data. The CCAR method explained earlier combines individual breaking segment attributes and aggregated attribute data in model estimation and prediction. It minimizes the potential bias with aggregated data mentioned above. There are limits to this approach. Aggregated attribute classes cannot be merged or divided between model estimation and prediction.

Modeling Scope Focus

Work in the model development phase included:

- Exploratory data analyses (histograms, correlations, outlier tests, plots, cross tabulations between the candidate predictor variables and observed break histories)
- Partitioning of the models by material and break type or level
- Considering discrete vs. continuous representation of predictor variables
- Considering linear and non-linear functional forms of the predicted breaks
- Considering multi-stage break models
- Statistical tests of "goodness of fit" of the models to the data
- Considering both individual and pooled utility data sets
- Validation tests of the models on separate subsets of the available data.

Results and elimination of less promising approaches

Elimination #1. Annual break probability for individual segments was estimated as a linear function of mains attributes and five years of cumulative breaks for Utility "A." The sample was about 4,000 individual segments. No aggregation of attributes was done. Installed year was removed as a predictor because it correlated highly with past breaks. Results were dominated by the break history variable, which was the only variable with high statistical significance. The R^2 measure of fit (under 0.1) was poor even for an individual segment probability model. This poor fit and general unavailability of individual segment data eliminated this individual break probability approach and turned the analyses toward segment aggregation.

Elimination #2. This model estimation aggregates the same attributes as the first one. An aggregated version of the same Utility A data was used to estimate the linear breaks model. Some 320 cross category cells were constructed as all combinations of 3 to 5 categories of 4 mains attributes and 2 categories of the past breaks. The 320 cells, with the average values of each attribute, were used in model estimation. The predicted variable for each cell was the number of breaks per year for all segments having that cell's combination of attributes. Each cell was weighted in the estimation by the proportion of mains having the attributes of that cell. Note many cells had zero predicted breaks and/or zero breaks history, since most main segments do not break. R^2 was better with this aggregation (R^2 = .64). This is because the prediction is for the average chance of breaking in the cells; it does not predict the probability of each segment breaking. R^2 is a measure of the accuracy of the average predictions, not individual segment breaks. Aggregation averages out unknown factors in individual breaks and random data error. The model makes it easier to see the relationships between the cell attributes and average breaks without these noise factors. If they have been carefully selected, homogeneous cross classifications of mains attributes and past breaks should make average predictions close to individual break probabilities.

The big difference between R^2 of the previous estimation and this aggregated one on the same data indicates significant unexplained factors in individual breaks, large random factors, data errors, or a combination of these. The aggregate model is unable to predict the variation in the individual breaks. This model was used for two further tests before a hybrid aggregated and individual data model was considered.

Elimination #3. Breaks in the Next 3 Years. One difficulty in individual segment predictions is predicting the year of failure. However, the year of failure is not as important as estimating which mains will fail soon. The team used the above model to predict probability of breaks in the next 3 years. The R^2 was better (.81 for three years versus .64 for the one year predictions). Thus breaks predicted over multiple years are more accurate. Thus the remaining development of models focused on predicting multi-year short term breaks.

Elimination #4. Joint Leak Estimations. A linear regression model of joint failures was also attempted. The results were weak compared to those for non-joint failures. R^2 for the initial one year and three year predictions were .21 and .35. The sample size of these breaks was several hundred. Because of the poor results and lack of interest and availability of joint leaks data from cooperating utilities, it was decided to drop joint failure predictions.

Elimination #5. Aggregation of Attributes. This effort showed practical problems with aggregated attribute categories. Some 320 cross classified cells were necessary to produce fairly uniform segment characteristics within the cells for 5 predictor variables. More cells would be needed to address unexplained factors. With a large number of cells, the more unusual combinations of attributes have very few segments. Small errors indicate erroneously large break rates in some cells. These produce outliers in the estimation data set unless many cells are checked and corrected. This time consuming and ad hoc process introduced additional errors. To address this, the team used the CCAR method to consider both non-breaking and breaking main data.

Break Types and time intervals

Break types were split/longitudinal, circumference, or small holes. Based on initial trials, three combinations were modeled: (1) rupture/blow-out/large hole, (2) the three other types combined, and (3) all four together. It was concluded that it was not worth the cost to predict two or more break type groups separately. Thus, all four break types were considered together in the modeling.

An ideal prediction would be of which segments would break each year. One year estimates are less reliable because of random effects or "noise." Models on data from several utilities were estimated for one, two, and three year periods of the predicted breaks. In almost all cases (different materials, model forms, etc.) the two year model gave significantly better results than one year predictions, and the three year predictions were not much better than two years. The remainder of the modeling was thus done for a two year prediction.

Functional Forms of the Models

The modeling approach was to use the historical pattern of breaks and attributes of a changing mains population to reveal relationships in the data rather than to impose a theory on the observations.

The team used linear functions of the predictor variables. It modeled individual segments using a CCAR database of mixed individual and aggregated data for segments that had broken in the 10 year window of data and those that had not broken. It used a first stage based on at least one break in any two years. A second model was used to predict the break count (per two years), conditioned on the chance of at least one break from the first stage model. Note this first stage prediction is not necessarily a first break as it is not possible to identify first breaks in most data.

This first stage prediction distinguishes the majority of segments with few or no breaks from the small percentage with break activity. The occasionally breaking segments require a second stage model. The first stage model also predicts the segments or segment classes that are not likely to break.

First stage models were estimated for two materials as linear regressions on the available mains attributes from three to four utilities. The measures of fit were better but still disappointing. As this prediction is a binary choice of break or no break, the logit function can be used.

For the Stage 2 multiple breaks model, linear and exponential functions (Weibull failure rate function) were used. Attempts to fit the Weibull function (of age and the other attributes) to the four-utility pooled data for a multi-breaks model gave poor fit or counter-intuitive results. This agrees with results for in the literature and showed that number of breaks, conditioned on at least one in 10 years, is not dependent on age of mains. Initial modeling also indicated that exponential functions of predictor variables for early breaks, similar to survival or proportional hazard models in the literature, did not fit the data as well as linear models.

Following these results, a linear form of multiple-breaks occurrences was chosen for the remaining estimations of the second stage model.

Development of the models

For model development estimates were based on pooled common data and on subsets of the data from the six individual utilities. The model effort involved three sets of prediction models for 3 mains materials.

The predictor variables considered were (number of indicator variables created for each factor by material type are shown in parentheses):

- Diameter groups (3-5)
- Installed year groups (3-6)
- Freezing degree-days, where applicable
- Annual rainfall
- Soil classifications (4-9), for some utilities only
- Joint types (4-5), for some utilities only
- Depth, for some utilities only.

Only the first four factors were common to data from all 6 utilities. Weather variables are not ordinarily predictors because they are not different between segments and are not known in the future. However, they help reduce unexplained variability in estimations. Joint type and depth were considered where available at two utilities. Limited soils data was available at one utility and publicly available soils data was related to geo-coded data at two others.

The first two attributes were used to define the cross-classified attribute records (CCAR). These two attributes enter into the models as the indicator variables for the same ranges of values that define their classes in the CCAR cells.

The soils and joint type attributes were also represented as indicator variables of attributes such as an installation period. They allow the attribute parameters to be interpreted as having either a continuous age influence on breaks or a discrete influence such as bad period in supply or installation.

The model results that follow are based on pooled data from the first 4 utilities that supplied data.

Stage 1 Results (Breaks or no breaks)

A logit function was used for the stage 1 probability of breaks. This functional form fit the data of one or more breaks better than a linear model. The predicted variable is the probability that a segment will have one or more breaks in a two year period. The observed values for the estimation are: $p = 1$ if a segment had one or more breaks in each two year period and $p = 0$ if no break occurred on the segment in that period.

The initial model estimations for each mains material were on data pooled from all utilities having that material and with no distinctions between break types predicted. The estimation sample was the breaking segments plus all non-breaking segments (or their cross-classified representation) over 8 years. As one example, the parameter estimates for steel mains had an R^2 of 0.267.

<u>Installed Year.</u> The parameter estimates for installed year did not vary in order of increasing age. Mains installed before 1905 and 1960-70 had a lower tendency to break than other years.

Diameter. The parameter estimates for diameter did increase for decrease in diameter as has been seen in other studies. Segments with 10 to 12 inch diameter had a lower tendency to break than other diameters.

Rainfall. For stage 1, steel breaks occurred more in lower rainfall years. Severe breaks, blowouts/ruptures, were not significantly related to rainfall. The utilities in our sample with steel mains were in coastal California and had no freezing degree-days. The rainfall variable had a consistent positive coefficient for the lined and unlined cast iron models predicting all non-blow/rupture break types. This appears to indicate that the less severe breaks (longitudinal, circumferential, and small holes) for these materials occur more often with higher rainfall, in contrast to steel. The significance levels of the rainfall variable in the models were high; they had minor but important influence on breaks for cast iron. The only significant relationship to the freezing degree-day variable was for lined cast iron breaks (other than blows/ruptures). The Stage 2, multiple breaks model showed poor statistical significance of the rainfall variable for all materials.

Past Breaks variables. The remaining two significant predictor variables are from the breaks history: sum of ruptures, large holes, or blow outs in the three years preceding the predicted breaks; and sum of all other types of breaks in the preceding three years. Both predictors showed the expected positive correlation between past and future breaks. It is also interesting to note the variable in the model that was not significant, sums of breaks in the past four to six years.

Soils data was directly available from only one of the six utilities. None of the others knew of local sources. Geo-coded STATSGO soil databases have been available free from the National Resources Conservation Service (NRCS) since 1994 for the lower 48 states. A finer geo-coded database for all counties is in preparation by the NRCS. When available soil measures were added to the models, several measures of significance were good, but they produced only a small improvement in predictions. For lined CIP soil measures that were significant predictors of breaks were water depth and unified soil class. For steel the steel corrosion index, water depth, and hydric index were significant. For unlined CIP the unified soil class was significant. The results indicate that the type of soil from coarse geo-data may not be very strong predictors of breaks compared to, for example, past breaks and mains diameter. But they can improve the predictions marginally and at low cost.

Multiple Breaks Models

In this modeling exercise, the predicted variable was the number of breaks in the next two year period. A linear function of the predictors was estimated.

An example of this model for unlined CIP with pooled data for 4 utilities and all break types used 1,475 break records. This included all segments that had at least one break in the sample interval (8 years for the typical utility). R^2 was only 0.042.

Only 3 of 11 inventory attributes were found to be significant predictors: the indicator for 6" mains, the indicator for mains installed between 1946 and 1952, and the amount of rainfall in the previous year.

Two historical variables from the 3-year type of breaks data files were also significant. Breaks occurring more than three years earlier had insignificant influences on future breaks.

Feasibility of Segment Level Predictions

To assess how well models predict at the segment level the team studied subdivided versions of its model validations. The results indicated small disagreement between actual and predicted breaks for groups of about 30 segments, but they show average deviations of 55% between predictions and observations at the individual segment level within the two year prediction period. There may be reason to consider an error factor as a conservative addition to the predictions.

Significance of Different Periods and Types of Break History

The team hypothesized that breaks from different past periods and different failure types can have different influences. It tested this theory and confirmed its intuition that recent major breaks are much stronger predictors of additional breaks than recent minor breaks.

Major breaks which occurred in the previous 4 - 6 years generally had only slightly more influence than those in the last 3 years.

Models by Material Type

Initially all models were estimated separately for four material types: unlined cast iron, lined or coated cast iron, steel, and asbestos concrete. The materials were separated because studies had shown differences between the behaviors of the materials. The team concluded that common models cannot be developed for two or more materials. Asbestos concrete was dropped from further analyses since only one cooperating utility supplied sufficient AC mains data.

Preliminary Conclusions on Modeling Methods

Conclusion 1. Neither totally aggregated nor totally individual data is appropriate for the data in the water community for the next decade or so. The hybrid CCAR method was chosen for the model estimations and predictions in this research and should be relevant for many years.

Conclusion #2. Multiple year predictions, with their better R^2 and more significant prior breaks variables were chosen for the model estimations. The utility databases were investigated before concluding to use 2 years as the prediction period.

Conclusion #3, No Joint Leak Models. Because of poor results and lack of interest and availability of joint leaks data it was decided to drop joint failure predictions.

MODEL APPLIED TO UTILITY DATA

Ideally, a model estimated from a representative data set from several utilities would predict breaks at other utilities. Often data on local factors such as construction and maintenance practices or soils are not available. The model parameters would vary between utilities if they were estimated independently.

This problem can be studied by validating a model estimated on pooled data against observed breaks of an individual utility. This can also be compared with a validation of a specific model estimation done for that utility.

Pooled data has much larger samples and different utilities have different segment or break counts. Thus, consistent comparisons between different utility samples or models require equal samples or adjusting results from each for sample size. Comparisons of predictions versus observations for probability predictions are more complex because the discrete variable has a much larger variance than continuous variables such as break count or probability.

The team compared the accuracy of its breaks/no-breaks model for unlined cast iron estimated on pooled utility data versus the same model on two utilities individual data. It concluded there are data errors in this model distorting the results and preventing clear conclusions.

In the second utility case, the team's results were conclusive on the direct application of break prediction models estimated on pooled data. They would make unacceptably large errors. It is only clear that the raw application of a pooled model is not acceptable.

A utility-specific model estimation would require about 200 hours of high skilled statistical labor, plus 80 hours for a QA technician, for a total cost of about $25,000 per utility.

SOME CONCLUSIONS ABOUT MODELING

Tests indicated that the type of break predicted (in contrast to the historical break types) did not produce significantly different models. Hence, all predictions were made for the average break type. However, the four types of historical breaks were found useful as predictor variables.

The best validated predictions were from a model to predict the probability that a main segment will break and a second model to predict the number of breaks. Their statistical product is the likely number of breaks on each segment.

The variables found to be the best predictors were:

- Past breaks
- Installed year
- Diameter
- Rainfall
- Other variables such as soil type, freezing degree days and joint type

A two year prediction model gave significantly better accuracy than one year predictions. Three year predictions were not much better than those for two years.

Prediction showed 55% deviation between predictions and observations for individual segments. Groups of 30 segments with similar attributes had 11% deviations. Therefore it is not as accurate to predict breaks on individual segments.

Significantly different model forms and coefficients were found by mains material. It was concluded that the materials had to have separate models.

Compared to utility-specific estimates, there was a 15 to 40% error when applying a model estimated on pooled data to an individual utility.

MAIN REPLACEMENTS JUSTIFICATION

By combining probability of failure with consequences, a planning model for main replacements is created.

The project was to produce break predictions and prioritized lists of segments justified for replacements. Segments were to be justified if the present value of their future breaks consequence costs being greater than the current cost to replace.

The inputs required were to be:

- At least 6 years of coded data on all past breaks showing their address or intersection locations, diameter, year installed, material, type of break and some other break qualifiers
- Categorical descriptions of mains mileage by diameter, year installed and material
- Unit costs of several breaks consequences (repairs, claims, etc.) and of replacements and by diameter (or acceptance of default national average costs)

Replacement of a main segment is justified when the present value of future costs maintaining the segment is greater than its cost of replacement. All direct and indirect consequences of breaks (property damage, water quality, loss of sensitive service, etc., as well as the cost of repairs) should be considered.

The costs are from utility-supplied data by diameter, type of break or other parameters. These are multiplied by the predicted breaks, discounting future year costs by the net interest on capital, to get the present cost of continuing to maintain a segment. The cost of replacement is the unit length cost of replacement times the length of segment replaced, plus the discounted annual cost to maintain the new main.

The sum of the calculated replacement costs for the justified segments is the recommended optimum budget for replacements in any year, providing a basis for justifying annual capital improvement budgets. Less capital expenditure will cause the utility to spend more in break consequences in the future. More replacements will also cause a greater total expenditure of maintenance and capital, particularly the latter.

The difference between the future costs of maintaining a segment and replacing it is defined as the "benefit" of replacement. Segments can be given normalized priorities for replacement by divided this benefit by their cost of immediate replacement.

Definitions of Mains Segments

Ideal data for a water distribution network would be records for each segment with unique attributes, such as diameter, year installed and location, and associated records describing all past breaks on that segment. Most water utilities have computerized break records but a minority of them have defined and coded their mains by street block and attributes. Fewer can directly link their break records to mains location. For a model of breaks prediction and replacement justification to be practical today, segments must be defined in terms of the available data for most utilities.

Replacement should also consider non-break reasons for replacements, such as public works coordination or mandates, network capacity factors or associated service lines and appurtenances. It also does not consider differential tax rates on expense versus capital funds nor depreciation factors.

CASE STUDIES

Case studies of the models were conducted at two utilities and partially tested at another. Introduction of the software system to the case study utilities required education on the definitions, reformatting, and corrections of breaks data. As mentioned above, a breaks format standard was intended as a common denominator for a system with many users. Many inconsistencies were found. The utilities were able to operate the software menus and non-breaks input and reporting screens easily, based on the User Manual.

The first, utility A, is a Midwestern municipal utility with nearly 100% unlined cast iron. The second, utility B, is a private West Coast water company having principally steel mains, plus cast and ductile iron and asbestos-concrete.

These utilities were among the six that initially supplied mains data to the research effort for the development of the breaks prediction models. Neither of these utilities had segment-defined mains data for their mains at the time of the study. Utility B had unique IDs for its break records, insuring against duplicates. The other utility did not.

Study descriptions

Each utility was supplied with the software and Users Manual for the prediction and prioritization system. Each copy was configured with the breaks model for one of their mains materials, loaded with the 8 to 12 years of the breaks records that they had supplied earlier, and with configuration parameters. The contact person in both cases was the manager of distribution maintenance or reliability, the same positions that had initially supplied their breaks data to the research project. The break prediction models put in the software for in each case were utility-specific versions estimated on their breaks data sets only.

Utilities were told they were expected to:

- Install the software from the CD
- Reformat their break records into the software format[10]
- Use the software loader and view its reports of any data problems
- Correct any inconsistencies, coding errors or rejections and reload
- Specify mains mileage by diameter and installed year on the supplied screen forms
- Accept system defaults or change unit costs for break consequences
- Accept system defaults or change unit costs for replacements
- When these were satisfactory, run the system prediction and benefit calculations and interpret the resulting Segment Break Prediction and Priorities report.
- Judge if the report gave reasonable replacement priorities and recommended budget.
- Fill out a questionnaire evaluating use of the software for their replacements strategies.

The utilities were also given quick response telephone support for the use of the software and a visit if necessary to help with its use.

The installation CDs for each utility also had one year of their previous breaks records already formatted into the REPLAN input standard so they could do test loads, try out the

[10] The software name was REPLAN for "Replacements Planning."

software, and obtain old priority reports with the system before they had to reformat their newer breaks.

The experience of these case studies follows, organized by the lessons it has for other users and changes that have or should be made.

Understanding and Usability

The first utility had problems with data preparation and formatting and with bugs in the program. Nearly 20 calls or emails and one visit for face-to face questions were required to give them useful results. Their greatest problems were with the required format and code definitions of the input breaks file. Most difficulty was because required break record fields were stored on 3 separate computers at this utility. Their unit costs data were also difficult to obtain. To get their segment break predictions and priorities report required about 5 person-weeks of effort over 4 months.

In the second utility the manager of the Service Delivery Group was able to use most of the system by referring only to the user manual. The first results on older data, without updating to latest breaks, recommended many of the segments they had earmarked or already replaced. The utility had improved their break location fields since previous data submissions, so they requested a modification that could be done with software settings for better data.

The software initially had a bug that prevented the user from getting necessary feedback on data formatting or coding problems in their new breaks loading. This was fixed and loading was done. The user was very interested in new features. Generally, the system gave this utility a tool they saw as helpful. This utility spent less than a person-week in formatting breaks and using the system.

Breaks Data Preparation and Loading

In the first utility, the difficulty supplying the breaks records was because data was in separate computers. Several of their numeric data fields were stored in character fields that did not convert automatically. Most field formats were different than the REPLAN input standard. There was also an initial problem defining the detailed material codes that belonged to a particular REPLAN material.

There were inconsistencies in their naming practices for street names. They found correcting this to be difficult or time-consuming. It is difficult for users to follow the address matching-type considerations that go into the model's proxy segments. Future users can be helped with a definition of the artificial segment ID that is created in this mode of operation. Inventory segment databases or GIS inventories would eliminate this problem but they will not be operational at most utilities for many years.

The breaks file format specification and methods for accommodating them were difficult to follow. More examples would help. Partly because their data came from several computers, they required a programmer proficient in Microsoft Access to accomplish the breaks file preparation Two person weeks were required but this yielded a software routine that would automatically produce the breaks input file for future updates of new breaks to REPLAN.

In the second utility, data preparation and reformatting was accomplished quickly, but the utility also felt the file format specification was too cryptic and needed more examples.

Prioritization report

The first utility saw that the interest rate they specified had to be supplied by their financial managers. Generally, experience with the gas industry has pointed to long range net interest rates at about 3%, independent of inflation. About 0.3% of the steel mains mileage analyzed was recommended for replacement by REPLAN for this utility. The net cost of capital they used for this was 3.75%. They had expected or hoped the system would justify more. Their mains are well maintained. This percentage of mileage is slightly smaller than that typically recommended by similar systems in the gas distribution industry unless the utility is behind in its replacements. The other utility had closer to 0.5% of its segments justified.

Case studies summary

Introduction of the system to the utilities required considerable education, support and experimentation with the data. One utility required much help in reformatting their data to the system standard, which was intended as a common denominator for many users, and correcting any break field errors or omissions. Many inconsistencies in the naming practices for street names were found. Correcting this was difficult or time-consuming. It was hard for a user to follow or accept the software proxy segments but they liked the results. Inputs of their mains mileage by diameter and installed year were an easy substitute for the absence of inventory databases. Inputting unit costs for consequences and replacements was easy. Getting accurate values for these costs and the appropriate net interest costs of capital within the utilities was much harder. The resulting segment priorities report appeared useful for their replacements planning and budgeting.

The areas of improvement suggested by the validation results are to develop methods to adjust pooled data models for application to individual utilities. It could demonstrate the cost tradeoffs between the four alternatives in that section: accepting the error level of pooled models for individual utilities; compensating for error range by conservative interpretation of the benefit priority that justifies replacement; incurring costs to develop utility-specific prediction models; or incurring development cost for the adjustment method and added utility data and staff effort to perform the adjustments.

Prediction accuracy could be improved by developing a Bayes statistical adjustment to all the models as has been done in the gas industry (Kulkarni, et. al. 1990).

Additional software features could be developed. One is to include edit capabilities for the segment priorities report. Other additions might be to add interfaces to other standards of breaks or mains attribute records, such as a GIS; adding tax and/or depreciation factors to the current interest rate provisions; adding cost factors for service lines or appurtenances that may be affected by replacements; considering public works coordination; and adding preventive maintenance/cathodic protection or other rehabilitation alternatives to be considered and prioritized or justified by the benefit/ cost calculations.

SUMMARY OF CHAPTER

This chapter has shown that data to support a breaks model is rarely complete without substantial data QA and transformations. Even where it is claimed to be available, data may take considerable quality checking, cleaning and processing for prediction use. Part of this (i.e., street name consistency or missing data and codes) may require manual effort by the utility. If the data has already been required to conform to another application such as in a GIS, the data may be much more suited to prediction models. GIS data may also require a standard segment definition algorithm for replacement.

The mains data processing and QA steps in this section, including equal street name resolution to identify multiple breaks on segments that are not keyed to mains segments, must be automated to produce a convenient system for continuing predictions of breaks. Considerable effort in this research has been devoted developing algorithms for this data processing automation.

Cause or condition variables were often available but are only useful for qualifying which type of a break is being predicted. Opinions about the weights of the underlying factors that are related to breaks are available. One utility with mains attributes by installed reach could not match their breaks to each reach but could identify both with aggregated zones containing tens of reaches. This allowed coarse but useful calculation of the attribute distributions of breaking versus non-breaking mains.

Preliminary findings on the data concluded that the analysis scope should be limited to above average break rate mains only (CIP, AC, and Steel, not DIP or plastic); and distribution mains only (large diameters and high pressures are special cases, and are too few per utility to reliably estimate statistical models).

CHAPTER 5
SUMMARY AND CONCLUSIONS

WHY UTILITIES NEED TO ANTICIPATE MAIN BREAKS

Water main breaks in the U.S. cost about $3 billion annually in direct costs and indirect consequences to critical infrastructure and traffic systems. This level of annual loss is on the same order of magnitude as all flood losses in the U.S.

Utilities require convincing business cases if they are to allocate funds for replacement and renewal of water mains. However, surveys and case studies show that it is difficult for utilities to prepare business cases and obtain funding for pipe renewal. Evidence includes the slow rate of water main replacement in the United States (once per 200 years or less often).

To overcome this problem utilities require more than just knowing they have an infrastructure problem. They must be able to anticipate pipe failures, both for classes of pipe in the aggregate and for individual water mains. Assessing risk of pipe failure in the aggregate helps with general planning but not to select individual pipes to renew.

Selection of individual pipes for renewal and replacement requires risk analysis using an effective main break prediction model, but no effective model is available in spite of several research studies. Types of main break prediction models can range from a utility's experience-based judgment as to which mains are at risk to sophisticated statistical models.

PROJECT BACKGROUND AND STATUS OF MAIN BREAK RESEARCH

AwwaRF anticipated the need for main break prediction models in the original project. Goals of the project were ambitious, including study of factors causing main breaks and the development and testing of software to predict main breaks.

AwwaRF's explanations of the need for the research remain valid ten years after the RFP was issued, but the context in which the research would be applied has changed. Now, utilities view their infrastructure problems as more serious than ten years ago, but researchers have shown that analytical methods can be developed but not verified or validated because the data bases are deficient.

This conclusion on the data bases recurs in research reports. Most recently, an AwwaRF report noted "Because of variations in the methods of reporting data and the difficulties in cross-correlation, it is recommended that an effective method of recording and organizing pipe asset and failure data be used, to capture this variation. Until data can be represented in this manner, care must be taken in interpreting and comparing failure rates" (Burn, et. al. 2005).

Utilities have known for years that they should assess condition and anticipate failure of water mains. Research into factors causing main began about 1980 and has focused on data management and analytical methods. In recent years, many utilities have begun to keep records on main breaks, at least in some form. However, datasets lack standard methods and quality control programs.

PRIORITIZATION OF WATER MAIN RENEWAL

Ideally, utility capital programs should show priorities that are based on probability of failure so that funds are applied to the most important investments. Any prioritization approach must fit within utility strategy and be credible to multiple groups. Although they cannot predict pipe failure with confidence, some utilities use point systems to assign priorities. This approach has merit, but is less satisfactory than knowing probability of failure, which will improve the preparation of business cases. Therefore, the ability to predict water main breaks and risks would enable utilities to use their capital funds better.

DATA MANAGEMENT

Data management requires standardization of break definition and protocols to collect and manage data. While emerging enterprise software systems help with data collection, standard methods are still missing. Until water main break data bases become more reliable and standardized, utilities will not have confidence in main break prediction models. Therefore, improving data protocols is the most urgent need to advance main break risk analysis.

MATHEMATICS OF MAIN BREAK HISTORIES

Several efforts have been made to model the statistical behavior of main breaks, and some of the results are consistent. One result is that different functions fit main break frequency early in the life of the main and later in the life of the mains. In other words, the time to the first break obeys one set of statistics, and the frequency of later breaks obeys another set. This seems logical because once a main begins to break, its problems are exposed. It is like a car; when it's new, it doesn't need many repairs. Once it starts to break down, it needs repairs more frequently.

Progress in modeling main breaks must be taken into perspective. Functions to describe early stage breaks are equations to predict how long a main will last before it begins to break. Analysis of later stage breaks has the potential to give utilities a tool to decide which mains to replace.

Mathematics of main break histories can be summarized with the statement that functional forms are understood, but lack of standard data prevents use of the functions at present.

RISK FACTORS FOR MAIN BREAKS

The link between main break models and risk factors is in the predictor variables that show risk of breaks. Examples are pipe age, diameter, material, soil corrosivity, and pressure, among others. General knowledge within the water industry shows that age of main is only a minor factor in the later stages of failure, which are most important for replacement decisions. Research has shown correlation coefficients for how breaks frequency depends on one or the other of the variables and sometimes how it depends on products of the variables.

If risk factors were the same across different utilities, then a generalized model could be developed. However, they are not the same, and pooled data cannot be used to predict main breaks. However, pooled data within a utility can replace missing data for individual main segments.

HUMAN FACTORS AND UTILITY ACCEPTANCE OF MODELS

The project showed that main break modeling is complex, and requires trained modelers. Utilities are wary of using "off-the-shelf" software until it is thoroughly proven, and the large investments required for pipe renewal require solid justification in budget processes, not predictions based on models without thorough testing and validation. Better ways must be found to explain main break data and models to managers and employees within utilities who would implement them.

OVERALL CONCLUSIONS AND FUTURE RESEARCH

The limits of model capabilities are not in mathematics or statistical analysis methods, but in lack of accurate and consistent data and in the complexity of analysis. Attempts to model main breaks use up-to-date analytical methods, and a number of mathematical models have been attempted, only to founder on lack of consistent data. Even if the model efforts were successful, they remain plagued both by data management and complexity of computational procedures.

Emphasis has been on modeling rather than data because modeling is easier and costs less. However, without adequate data, model efforts may not be useful. While main break research has been ongoing for over twenty years, progress has been slow. Rather than being a cause for discouragement, that slow progress signals a need to keep on with the research. The focus in the next stages of research should be on the datasets to predict main failures.

CHAPTER 6
RECOMMENDATIONS TO UTILITIES AND AwwaRF

UTILITY RISK ANALYSIS IN REPLACEMENT PLANNING

It is widely understood that utilities have an infrastructure problem and that buried assets are the major share of utility infrastructure. However, the renewal rate of aging and sometimes failing infrastructure is low, and this low rate can be explained by a risk-averse approach to decision making, lack of data to prove the need for more aggressive approaches, and a desire by utility management to keep rate increases low and in the politically-acceptable range.

To achieve improved renewal rates, utilities should perform risk analysis to anticipate main breaks so they can prevent direct losses to their own finances and infrastructure, and they can alleviate customer and public losses. To know how to do this, they require convincing business cases for pipe renewal, which require that they anticipate failures of individual water mains. This supports capital programs in which funds are applied to the most important investments and utility strategy is credible to customers and regulators. To be able to predict failure requires an effective main break prediction model, even if the model is not quantitative but is a judgment–based assessment by experienced utility personnel.

HOW UTILITIES CAN PERFORM RISK ANALYSIS FOR MAIN BREAKS

Utility risk analysis is required for capital programs, so utilities should strive to gain the capability to analyze main breaks. Larger utilities do this now, so a major issue is to develop methods that work for smaller utilities.

If a utility chooses to study its main break rate frequencies, it can use the research in this report to inform its engineers and researchers about the experience of other utilities in predicting break rates. This utility would then need to develop and test methods for application to its own system.

While the industry knows intuitively that some risk factors are common across utilities, general knowledge within the industry shows that age of main is only a minor factor in the later stages of failure, which are most important for replacement decisions. The influence of other risk factors, such as material, soils, corrosivity, temperature, and loading, require further study in individual utilities. Recent rates of main breaks and consequences of failure drive decisions as much as analysis of these risk factors.

Having a reliable main break risk model would greatly aid utility capital planning. However, the project showed that main break modeling is complex, and requires trained modelers, even if the data is available to calibrate models.

Currently, no proven off–the–shelf models are available to predict break rate frequencies, but several models or model concepts are available for utilities to adapt in their planning efforts. For example, Deb et. al. (2002) offer a procedure for ranking pipes for replacement. This approach, along with others such as the one explained by Hope (2003), offer a starting point for utility applications.

Utilities usually use judgment to consider multiple main break variables. Often, renewal plans are reactive or based on opportunities, such as replacing seemingly problematic pipe when streets are re-paved.

Until main break prediction improves, utilities can develop replacement models based on experience-based point scores. Some leading utilities use this method, which in fact incorporates the risk factors identified by the utilities. This is a straightforward approach to capital planning, and utilities can use methods outlined in the Capital Planning Strategy Model (CH2M Hill. 2002) for their decision analysis. This approach has merit, but is less satisfactory than knowing probability of failure to strengthen business cases for pipe renewal.

To pinpoint their needs for main break models, utilities can identify the exact decisions that depend on knowing risk of main failure. Generally speaking, these are to allocate funds for renewal in capital improvement planning and to schedule rehabilitation and/or replacement in individual pipes. The capital allocation can be done without modeling individual mains because utilities can track the number of main breaks across the network and they can assess direct and indirect costs of main breaks to determine return on investment of renewal. The more difficult problem is to select individual mains for renewal. Here the analogy is more like a medical problem than like replacing vehicles in fleet or selecting road segments for renewal. The medical problem and the main break problem both involve potential disasters, failure of a human body part and maybe death or a main break that causes a high level of damage. Therefore, it pays to invest additional funds to determine specifically which main (or body part) is most at risk to fail. This risk assessment inevitably involves data and modeling, but the modeling can be done without necessarily assigning mathematical functions to the data. It can also be done to sort mains by risk factors using simple methods. The key is a consistent and coherent data base.

Utilities can develop learning methods to identify risk factors, and for each new break they can add new lessons to their data bases. They also can train personnel for analysis of distribution systems. They can initiate data collection and analysis programs, including main breaks research programs based on data.

WATER SUPPLY INDUSTRY RESEARCH

Utilities view infrastructure problems as more serious than before, and AwwaRF has ongoing research programs about infrastructure reliability. AwwaRF anticipated the need for main break prediction models in the original project and its explanations of the need for the research are still valid.

While main break research has been ongoing for over twenty years, progress has been slow. Rather than being a cause for discouragement, that slow progress signals a need to keep on with the research. Emphasis has been on modeling rather than data because modeling is easier and costs less. Researchers have shown that analytical methods are available but data bases are deficient. Without adequate data, further model efforts may not be useful. Therefore, the focus in the next stages of research should be on datasets to predict main failures.

In addition to this project another recent AwwaRF report noted the same need: "Because of variations in the methods of reporting data and the difficulties in cross-correlation, it is recommended that an effective method of recording and organizing pipe asset and failure data be used, to capture this variation. Until data can be represented in this manner, care must be taken in interpreting and comparing failure rates" (Burn, et. al. 2005).

The further research required would focus on the problem that datasets lack standard methods and quality control programs. Data management requires standardization of break definition and protocols to collect and manage data. While emerging enterprise software systems help with data collection, standard methods are still missing. Until water main break data bases become more reliable and standardized, utilities will not have confidence in main break prediction models. Therefore, improving data protocols is the most urgent need to advance main break risk analysis.

After protocols for data management are developed, the further issue to be tackled is how to create incentives for utilities to collect and manage the data and to overcome institutional obstacles that prevent better data management systems. In addition, to recognize the complexity of analysis, better ways must be found to explain main break data and models to managers and employees within utilities who would implement them.

APPENDIX
SURVEY RESULTS AND QUESTIONS

EXPLANATION

This appendix has a summary of the results and questions from the original project survey. It has been prepared from the project's draft report (URS 2002). Some 45 utilities responded, with an average of 164,571 service connections. See Chapter 3 for a discussion of the results.

SURVEY QUESTIONS

This is a summary of the questions asked in the survey instrument used in the original project. No attempt has been made to retain the format of the original questionnaire, but the substance of the questions asked is summarized below.

Question	
1	General information about distribution system Main inventory by material, no. of services, ownership, no. of failures, material with worst failure record
2a	Method to plan replacements considering breaks and leaks Replace only most serious failures Experienced opinion on mains needing replacement Categorical rule (> 40 years old, > 3 breaks, other) Simple model (e.g. demerit points for prior breaks, pressure, risk area, etc.) Rating model from the literature
2b	Other methods to plan replacements Hydraulic grid capacity criteria or model Water quality regulations or criteria Fire protection criteria Transmission line or other facility constraints Lists of public works (e.g. paving) coordination or opportunities Other
3	Problems in selecting break-related replacements (Order by importance, 1, 2, 3, etc.) Estimating life expectancy or failure chances of mains sections Estimating costs of future main repairs and replacements Evaluating the consequences of main leaks and breaks Focusing on particular main segments to replace Justifying or supporting replacement costs/budgets (to management, city, or PUC) Resource limitations for evaluating mains (data, staff skills, etc.) Other

4a	Implementation of method or model of breaks prediction; main lifetimes, or replacements planning from the literature (such as those below) AwwaRF Pub. 490509 "Water Mains Evaluation for Rehabilitation/Replacement", O'Day. 1986 AWWA Journal, "An Analytic Approach to Scheduling Pipe Replacement" Shamir and Howard. 1979 EPA Study "Determinants and Options for Water Distribution System Management", Clark, 1982 M.I.T. Study "Statistical Models for Water Mains Failures" (failure models based on New Haven and Cincinnati Water Dept. data). Marks. et al. EPA Report 60015-871003.1987 Your own method or model Other reference used
4b	Success in implementing a formal replacements model or plan If Yes, describe its use, benefits (or your own model) If no, why were you not successful? Data limits, staff skills, cost to implement, too complex, other?
5	Mains data availability Mains failure records (structural or joint failures) – yes or no answers General description (date, address, time, etc.) Main description (material, diameter) Pressure (typical, pump status, etc.) Other main description (year installed, depth) Indicate if it was a break (large, emergency, etc.) if it was a circumference break if it was a blow out (piece of pipe wall) if it was a longitudinal split if it was a split bell/joint if it was a break from soil movement Indicate if it was a leak (small/schedulable) if it was a joint leak if it was a local hole. external corrosion if it was a local hole, internal corrosion Visual observations (pipe &joint condition nearby construction, traffic load, etc.) Bedding condition Water temperature at failure time Geotechnical/seismic factors (soil, slope/site. etc.) Do you keep field samples? Type of repair or replacement done Indicate if a failure due to 3rd party hits Indicate if a planned, not failure-caused action Are the items you answered Yes to in a database If yes, how many years of failures are in your computer database If no, how many years of paper records to you have? Do you plan a database?
5b	Mains/material inventory

	Do you have mapping or GIS software containing all your mains attributes? Mains data only on maps or drawings If mains data in software, a database, or tables, what are the smallest subunits for which you have data? (blocks, reach, zone, all mains together) For your smallest sub-units of mains, do you record the following attributes for each unit? (Yes/no) Material ? Diameter? Year installed? Length of reach, segment, or subarea? Depth of cover? Corrosion protection? Joint types? Operating pressure? Soils or resistivity data for whole area? Soils data only where breaks occurred? Which computer tools do you currently use for mains inventory or breaks data, or plan to install within 5 years? Geographic Information Systems Portable or hand-held field computers SCADA system Global Positioning System (GPS)
6	Consequences of failures Which consequences of breaks or leaks are important to you (order by severity) Damage claims, lost revenue claims, replacement/rehab costs, repair costs, fire protection, injury claims, customer-hrs/outage, water quality, lost water cost, traffic control cost, public confidence
7	Availability of Consequence Data Do you maintain any of these types of data for evaluating local breaks consequences? Property values, population, sensitive land uses, traffic volumes Other breaks consequence data?
8a	Project serve your needs That is, assuming you have or could develop the required data, would you like to obtain a validated breaks prediction & mains replacement prioritization software package (based on your mains data and an accounting of your cost and risk consequences)? Why? Breaks leaks impacts? To justify replacement budgets? Repair costs?
8b	Problems with project Data limits, staff skills, cost to implement, other resources.
9	Preference for model Would you prefer an individual segment replacement priority system or a proven categorical rule (e.g. all pipes > 40 years old and < 8" diameter and in high impact areas. (Note an individual segment model has greater data requirements and effort to

	implement. but lower costs of replacements because of its segment focus. A categorical rule is easy to implement but due to its broad categories will suggest replacing many good main segments along with the bad ones.)

Survey results

1. Mains inventory and failures

	Percent of miles	Average miles
CI	48%	926
AC	11%	212
DI	22%	432
Plastics	13%	254
Steel	6%	111
Other	0.1%	2.7
Total miles	100%	1,938

2a. Selection method for replacement (could give multiple responses)

Method	% of responses
Serious breaks only	28.6
Expert opinion	50.0
Specific categories	35.7
Points-rating system	19.0
Published model	2.4

2b. Importance of other factors versus break prediction
(1 = most to 6 = least)

Factor	Importance
Break prediction	2.0
Capacity	3.2
Quality of water	3.5
Fire protection	2.6
Transmission constraints	3.7
Public works coordination	3.3

3. Problems in selecting mains for replacement
(1 = most to 6 = least)

Factor	Importance
Assess mains life	2.6
Assess costs	3.4
Assess consequences	2.9
Focus on segments	2.8
Justify replacements	2.7
Resource limits	2.9

4a. Model use from literature (could give multiple responses)

Method	% of responses
Tried one from literature	27.9
AwwaRF Report 90509	11.6
Shamir and Howard, 1979	4.7
Clark '82	0
MIT '87	0
Corps of Engineers '82	0
Own method	20.9
Other	4.7

4b. Success implementing model

Success level or problem	% of responses
Success?	32.6
Data limits	31.3
Skills	25.0
Costs	25.0
Complexity	25.0
Other	43.8

5a. Data recorded

Success level or problem	% of responses
date, address, time	97.6
material, diameter	97.6
Pressure	31.0
year installed, depth	67.5
if a break	84.2
if a circumferential	85.7
if a blow out	73.8
if a longitudinal split	78.6
if a split bell	78.6
if due to soil movement	52.4
if a leak	75.0
if a joint leak	78.0
if external corrosion	56.1
if internal corrosion	47.5
Visual observations	53.7
Bedding condition	35.7
Water temperature	4.8
Geotechnical/seismic factors	26.2
keep field samples	16.7
repair method	81.0
if 3rd party	82.9
planned repair, no failure	48.8
in a database	51.2
years in database	14.5 years

5b. Mains inventory basics

Data	% of responses
All in map software	46.2
On maps only	66.7
By segment	27.9
By reach	11.6
By zone	0
Aggregate info only	7.0

5c. Computer technology

Data	% of responses
GIS now	41.9
GIS planned	39.5
Field computers now	9.3
Field computers planned	39.5
SCADA	30.2
SCADA planned	14.0
GPS now	11.6
GPS planned	25.6

6. Importance of consequence factors
(1 = most to 11 = least)

Consequence factors	Score
Damage claims	3.9
Lost revenue claims	7.0
Replacement costs	6.1
Repair costs	3.7
Fire protection	3.2
Potential injury	5.2
Outage hours	2.5
Quality of water	3.2
Lost water	7.8
Traffic	7.1
Confidence	4.0

7. Availability of consequence data (could give multiple responses)

Data	% of responses
Property values	7.0
Population densities	7.0
Sensitive uses	7.0
Traffic volumes	4.7
Other available data	7.0

8a. Project serve needs

Question	% of reponses
Project does serve needs	72.1
Why, assess impacts?	4.8
Why, justify replacements?	60.5
Why, predict repair costs?	39.5

8b. Reason project of no use (could give multiple responses)

Reason	% of responses
Data limits	9.3
Skills	7.0
Costs	9.3
Other resources	2.3
Want advice only	9.3

9. Preferred level of prioritization detail

Level	% of responses
Individual segments	67.4
Categorical on factors	16.3

REFERENCES

Agbenowosi, Newland. 2000. *A Mechanistic Analysis Based Decision Support System for Scheduling Optimal Pipeline Replacement*. Ph. D. Dissertation, Virginia Polytechnic Institute and State University. Blacksburg, VA.

Andreou, S. A., Marks, D.H., and Clark, R.M. 1987b. A new methodology for modeling break failure patterns in deteriorating water distribution systems: Applications. *Advances in Water Resources. 10(1)11-20.* Elsevier Science, B.V.

Andreou, S.A. 1986. *Predictive models for pipe break failures and their implications on maintenance planning strategies for deteriorating water distribution systems.* Ph. D. Dissertation. Massachusetts Institute of Technology. Cambridge, MA.

Andreou, S.A., D.H. Marks, and R.M. Clark. 1987a. A new Methodology for Modeling Break Failure Patterns in Deteriorating Water Distribution Systems: Theory. *Advances in Water Resources. 10(1):2-10.* Elsevier Science, B.V.

Arnold, G.E., E.J. Clark, G.J. Remus, and H.W. Niemeyer. 1960. Experience with Main Breaks in Four Large Cities. *Jour. AWWA,* 52(8):1041-1058.

AWWA. 2003. *Water://Stats 2002 Distribution Survey.* Denver.

AwwaRF. 1997. *Main Break Prediction, Prevention, and Control.* RFP 461. Denver, CO.: AwwaRF.

Baracos, A., W.D. Hurst, and R.F. Legget. 1955. Effects of Physical Environment on Cast–Iron Pipe. *Jour, AWWA*, 47(12):1195-1206.

Boynton, P.G. 1989. *Analysis of Corrosion Related Main Break Data.* Unpublished Paper obtained from L. D. McMullen, DesMoines Waterworks, DesMoines IA.

Burn, S., Davis, P., Schiller, T., Tiganis, B., Tjandraatmadja, G., Cardy, M., Gould, S., Sadler, P., Whittle, A. 2005. *Long Term Performance Prediction for PVC* Pipes. Denver, Colo.: AwwaRF and AWWA.

CH2M Hill. 2002. *Capital Planning Strategy Manual.* Denver, CO.: AwwaRF and AWWA.

Clark, R. M., C. L. Stafford, and J.A. Goodrich. 1982. Water Distribution Systems: A Spatial and Cost Evaluation. *Jour. Water Resources Planning and Management Division, ASCE,* 108(WR3) 243-256.

Cromwell, J., G. Nestel, R. Albani, L. Paralez, A. Deb, and F. Grablutz. 2001. *Financial and Economic Optimization of Water Main Replacement Programs.* Denver, Colo.: AwwaRF and AWWA. .

Cromwell, J., H. Reynolds, and K. Young. 2002. *Costs of Infrastructure Failure.* Report #90918. Denver, Colo.: AwwaRF and AWWA.

Cullinane, M. J. Jr., J. Goodrich, and I. Goulter. 1989. Water Distribution System Evaluation. In *Reliability Analysis of Water Distribution Systems.* Edited by L. W. Mays. New York, NY: American Society of Civil Engineers.

Damodaran, N., J. Pratt, J. Cromwell, J. Lazo, E. David, R. Raucher, C. Herrick, E. Rambo, A. Deb, and J. Snyder. 2005. *Customer Acceptance of Water Main Structural Reliability.* Denver, Colo.: AwwaRF and AWWA.

Deb, A.K., K.A. Momberger, Y.J. Hasit, and F.M. Grablutz. 2000. *Guidance for Management of Distribution System Operation and Maintenance.* Denver, Colo.: AwwaRF and AWWA.

Deb, A.K., Y.J. Hasit, F.M. Grablutz, and Herz, R.K. 1998. *Quantifying Future Rehabilitation and Replacement Needs of Water Mains.* Denver, Colo.: AwwaRF and AWWA.

Deb, A.K., Y.J. Hasit, H.M. Schoser, and J.K. Snyder. 2002. *Decision Support System for Distribution System Piping Renewal*. Denver, Colo.:AwwaRF and AWWA.

Dolson, F.E. 1955. Factors Causing Main Failures. *Jour. AWWA,* 47(2):465-469.

Elnaboulsi, J. and Q. Alexandre. 1997. Determination de la date optimale de renouvellement des reseaux urbains d'eau potable. *Canadian Water Resources Journal.* Cambridge, CANADA, 22(3):327-350.

Federal Emergency Management Agency. 2006. Flood Statistics. [online]. Available: <http://www.floodsmart.gov/floodsmart/pages/statistics.jsp.> [Cited June 28, 2006]

Goodrich, J, C.P.S. Frebis, E. J. Read, and Robert Clark. 1987. Management of Water Distribution Systems Components. *Proc. National Conference on Hydraulic* Engineering. Williamsburg, VA.

Goodrich, J., L.W. Mays, Y Su and J Woodburn. 1989. Data Base Management Systems. In *Reliability Analysis of Water Distribution Systems.,* Edited by L. W. Mays. New York, NY: American Society of Civil Engineers.

Grigg, Neil S. 2005. Assessment and Renewal of Water Distribution Systems. *Jour, AWWA*, 97(2):58-68.

Grigg. Neil S. 2004. *Water Distribution Assessment and Renewal.* Denver, Colo.: AwwaRF and AWWA.

Hope, Richard. 2003. *Establishing the Estimate of Annual Replacement Rate for Water Mains.* Workshop for Infrastructure: Above and Below Ground. AWWA ACE 2003, Anaheim.

Horn, Tom. 2003. *Small Line Replacement and Prioritization, Columbus Water Works.* Presented at Distribution System Infrastructure Integrity Workshop. April 3-4. Fort Collins, CO: Colorado State University.

Hosoi, Y., H. Murakami, M. Kozai, and S. Marutaka. 1989. Consideration of Water Main Breakage *Japan Water Works Association, 58(5):* 10-16.

Kane, M. J. and A. Gough 1996. Database to Prioritise Mains Rehabilitation. (Corporate author: Severn Trent Water Ltd., Birmingham, England). *IWSA Specialized Conference on New developments in Modeling Monitoring and Control of Water Supply Systems*, Amsterdam, Netherlands, September 24 -5, 1996.

Keeley, Kurt. 2003. *AWWA Survey of Operating Data, 1945-1985.* Unpublished document. AWWA. Denver.

Kettler, A. J. and I. C. Goulter. 1985. An Analysis of Pipe Breakage in Urban Water Distribution Networks. *Canadian Journal of Civil Engineering,* 12: 286-293.

Kirmeyer, G.J., W. Richards, and C.D. Smith. 1994. *An Assessment of Water Distribution Systems and Associated Research Needs.* Denver, CO: Denver, Colo.: AwwaRF and AWWA.

Kleiner, Y. B. Rajani, and R. Sadiq. 2005. *Risk management of large diameter water transmission mains*. Denver, Colo.: AwwaRF and AWWA.

Kleiner, Y., Sadiq, R., and Rajani, B. 2004. Modeling Failure Risk in Buried Pipes Using Fuzzy Markov Deterioration Process. *ASCE International Conference on Pipeline Engineering and Construction.* San Diego. August 2-4, 2004.

Kleiner, Yehuda and Rajani, Balvant. 2001. Comprehensive review of structural deterioration of water mains: statistical methods. *Urban Water,* 3, 131-150.

Kulkarni, R. and F. Reid. 1992. *Analytical Techniques for Selection of Repair and Replace Options for Cast Iron Piping Systems.* Final Report, Contract GRI-9210234. Chicago, IL: Gas Research Institute (GRI).

Kulkarni, R. B., K. Golabi, G. Dugovic and K. Burnham. 1990. Optimal Replacement decisions in Cast Iron Piping Systems. In *Prioritizing and Pipe Replacement Decisions,* Technical Report No XL-8920, American Gas Association, 34-39.

Lim, E. L.1997. *Pipe Evaluation System (PIPES).* Environmental Systems Research Institute (ESRI) User's Conference, San Diego, CA, July, 1997.

Loganathan, G.V., Sherali, H.D., Park, S., and Subramanian, S. (2002). *Optimal Design-Rehabilitation Strategies for Reliable Water Distribution Systems.* Virginia Polytechnic Institute and State University, Water Resources Research Center. Blacksburg, Virginia.

Madioc, H., P. Botzung. B. Bremond and P. Eisenbeis. (Lyonnaise des Eaux, Bordeaux, France). 1996. Diagnosis Methods and Performance Indicators for Rehabilitation Policies. *Water Supply*, 12(314)347-358.

Marks, D H., S. Andreou, C. Jeffrey, D. Park and A. Zaslavsky. 1987. *Statistical Models for Water Mains Failures.* Contract CR810558, U.S. E.P.A. Drinking Water Research Division, Water Engineering Research Laboratory, Report 1. EPA/16015-871003108, Cincinnati, OH, 108 pages.

Mavrakis, George. 2003. *Evaluating Condition of Water Facilities Using Predictive Life Cycles.* Presented at Distribution System Infrastructure Integrity Workshop. April 3-4. Fort Collins, CO: Colorado State University.

Mays, L.W. 1990. Models of Optimal Maintenance of Hydraulic Structures. In *Proc. of 5ymposium: Water Resource infrastructure: Needs, Economics, and Financing.* J.F. Scott and R.M. Khanbilvardi, Ed. Ft. Worth, TX: NY: American Society of Civil Engineers.

Morris Jr., R.E. 1967. Principal Causes and Remedies of Water Main Breaks. *Jour. AWWA,* 59(7):782-798.

O'Day, D.K., R. Weiss, S. Chiavari, and D. Blair. 1986. *Water Main Evaluation for Rehabilitation/Replacement.* Denver, Colo.: AwwaRF, AWWA, and Environmental Protection Agency.

Parmelee, Mary. 2006. Water industry's optimism eroding. *AWWA Mainstream.* Vol. 3, No. 13.

Pelletier, Genevieve, Mailhot, Alain, and Villeneuve, Jean Pierre. (2003). "Modeling Water Pipe Breaks – Three Case Studies." *Journal of Water Resources Planning and Management*, 129 (2), 115-123.

Rogers, Peter. 2006. *Failure Assessment Model to Prioritize Pipe Replacement in Water Utility Asset Management.* Ph.D. Dissertation, Colorado State University. Fort Collins, Colorado.

Runge, J. and Mann, J. 2005. State of the Industry Report, 2005. *Jour, AWWA,* 97(10):58-67.

Seargeant, Doug. 2003. Water Main Rehabilitation: Prioritization. *ACE Convention.* Anaheim, CA.

SEAMS. 2002. *Decision Support for Clean Water Infrastructural Planning in UUplc.* Sheffield, U.K.

Shamir, U. and C.D.D. Howard. 1979. An Analytic Approach to Scheduling Pipe Replacement. *Jour. AWWA*, May, 1979.

Thames Water Utilities. 1998. *Utilnets: model for hydraulic impact consequences of mains bursts.* London, U.K.

URS Corporation. 2002. *Water Main Breaks Prediction, Prevention and Control.* Draft Final Report. Oakland, CA.

Vanrenterghem, Annie. 2003. *Modeling the Structural Degradation of an Urban Water Distribution System.* Ph. D. Dissertation, Polytechnic University. Brooklyn, N.Y.

Walski, T. M. and A. Pelliccia. 1982. Economic Analysis of Water Mains. *Jour. AWWA,* 74(3):140-147.

Wolfe, T. F. 1946. How to Prevent Breaks in Cast-Iron Pipe. *Jour, AWWA,* 38(6)765.

Woodward-Clyde and San Jose Water Company. 1997. *Main Break Prediction, Prevention, and Control*. Proposal to AwwaRF. Oakland, Calif.

ABBREVIATIONS

AC	Asbestos Cement Pipe
AWWA	American Water Works Association
AwwaRF	Awwa Research Foundation
CCAR	Hybrid cross-classified attribute record
CIMOS	Cast Iron Maintenance Optimization System
CIP	Cast iron pipe
DIP	Ductile iron pipe
DMA	District Meter Area
EPA	Environmental Protection Agency
GIS	Geographic Information System
ID	Identification number
KANEW	Assessment tool ("Karlsruhe New")
LADWP	Los Angeles Department of Water and Power
MIT	Massachusetts Institute of Technology
NRCS	National Resources Conservation Service
PAN	Priority action number
PIMOS	Pipe Maintenance Optimization System
PIPES	Seattle's mains evaluation system
PREDIKASS	Lyonnaise des Eaux's pipe assessment tool
PVC	Polyvinyl chloride pipe
QA/QC	Quality assurance and control
R^2	Coefficient of determination
REPLAN	Software tool for replacement planning
RFP	Request for proposal
RFP	Request for proposal
SDWA	Safe Drinking Water Act
Utilnets	A tool of Thames Water Utilities, Ltd.
WATER://STATS	AWWA's water utility database

IWA Publishing's authorised EU representative for General Product Safety Regulations is Diane D'Arras, 15 rue Duret, 75116 Paris, France, e-mail: safety@iwap.co.uk.

Printed and bound by CPI Group (UK) Ltd, Croydon, CR0 4YY
17/09/2026
02218793-0014